Noise and Vibration Control in Buildings

ROBERT S. JONES

Sponsored by Bolt Beranek and Newman Inc.

McGRAW-HILL BOOK COMPANY

New York St. Louis San Francisco Auckland Bogotá
Hamburg Johannesburg London Madrid Mexico
Montreal New Delhi Panama Paris São Paulo
Singapore Sydney Tokyo Toronto

DEDICATED TO

My mom and dad, for their love and a good home,
My wife Jo, who has always been my support,
My three sons, Christopher, Bradford (deceased), and Eric,
of whom I am very proud, and my beloved grandson
Tyler Buckminster Jones.

Library of Congress Cataloging in Publication Data

Jones, Robert S. (Robert St. Clair), date.
  Noise and vibration control in buildings.

  Bibliography: p.
  1. Soundproofing.   2. Buildings—Vibration.   I. Title.
TH1725.J58   1984   693.8′34   83-24855
ISBN 0-07-006431-8

1234567890   DOC/DOC   8987654

ISBN 0-07-006431-8

The editors for this book were Joan Zseleczky, Geraldine Fahey, and Susan West,
the designer was Naomi Auerbach, and the production supervisor
was Thomas G. Kowalczyk. It was set in Melior
by Progressive Typographers.

Printed and bound by R. R. Donnelley.

Contents

Foreword   xiii
Preface   xv
Acknowledgments   xvii

1. Introduction   **1**

   Structure of the Manual   **2**
   Key Concepts Used in Noise and Vibration Control   **2**
   Key Terms Used in Noise and Vibration Control   **3**

**2. Guidelines for Architects, Engineers, and
Contractors: Vibration Isolation and Noise
Control Treatments**   **7**

   General Information   **8**
     Quality Assurance   **8**
     Submittals   **8**
   Products   **9**
     Floor Isolators   **9**
     Hanger Isolators   **10**
     Wall Sleeves   **11**
     Equipment Bases   **11**
     Vibration-Isolation Curbs   **12**
     Exterior Metal Parts   **12**
     Equipment Mountings   **13**
     Duct Silencers   **13**
     Acoustical Duct Lining   **14**
     Acoustical Sealant   **15**
   Execution   **15**
     Installation of Vibration-Isolation Devices   **15**
     Resilient Support of Piping   **15**

Connections of Piping 15

Resilient Support of Ducts 16

Connections of Ducts 16

Operation of Rotating Equipment 16

Installation of Duct Silencers 16

Penetrations of a Structure 16

3. Mechanical Systems: Heating, Ventilating, Air Conditioning (HVAC)   19

VIBRATION ISOLATORS 21

Hangers 21

Vibration-Isolation Hanger 21

Vibration-Isolation Unit 23

Vibration-Isolation Hanger for Air-Handling Unit 24

Vibration-Isolation Hangers 25

Vibration-Isolation Hangers 26

Housed Spring Isolators 28

Housed Vibration Isolator 28

Housed Spring Isolators 30

Housed Vibration Isolator 31

Housed Isolators 32

Isolators 33

Housed Isolator with Multiple Springs 35

Unhoused Spring Isolators 35

Open Spring Isolator 35

Single-Spring Vibration-Isolation Unit 36

Double-Spring Vibration Unit 38

Double-Spring Isolator 39

Spring Isolator for Cooling Tower 40

Vibration Isolator with Travel Limit Stops 40

Spring Isolator with Unreleased Travel Limit Stops 41

Spring Isolator for Fan 42

Spring Isolator with Lateral Restraint 43

Misaligned Spring Isolators 43

Unhoused Spring Isolator 44

Vibration-Isolated Pump 46

Isolator for Air-Handling Unit 47

Floor Spring Isolator for Air-Handling Unit 48

Isolator for Rooftop Equipment 49

Floor-Supported Equipment 49

Floor Vibration Isolator 50

Floor-Mounted Pipe Isolator 51

Floor-Supported Piping Isolation 52

Air Spring 52

Pneumatic Mounts 53

Air Springs and Mounts 55

DUCTS 57

Silencers and Enclosures 57

Duct Silencers 57

External Duct Treatment 59

Preferred Duct Silencer Location 61

Incorrect Installations of Duct Silencers 63

Correct Installations of Duct Silencers 64

Duct Silencer Connection Installation 66

Duct Silencer Acoustics Test Setup 67

Duct Penetrations and Connections 70

Duct Penetrating Wall 70

Oversized Duct Penetration 71

Exterior Rain Hood at Flexible Duct Connection 72

Horizontal Thrust Restraint 74

Flexible Duct Connection 74

Air-Conditioning Duct for Noise-Sensitive Spaces 75

Duct Shafts and Steel Angle Supports 76

Duct Horizontal Supports 77

Internal Acoustical Treatment of Air Duct and Access Doors 78

PIPES 80

Single Pipes 80

Single-Pipe Isolation 80

Condenser Water Main from Cooling Tower 81

Pipe Penetrating Building Structure 82

Condenser Waterline with Fireproofed Hangers 84

Vibration Isolators for Condenser Water Pipes 84

Condenser Water Piping Isolators 85

Pipe Hanger Embedded in Building Wall 86

Heat-Pump-System Water Main with Isolators in Wall 87

Floor-Mounted Vibration Isolation for Piping, Chiller
Equipment Room 88

Pipe Penetration Sealing Device 90

Vibration-Isolation Hanger 90

Vibration-Isolation Hanger 91

Vibration-Isolation Hanger Rods 91

Vibration Isolation 92

Vibration-Isolation Hangers   92

Roller-Type Hanger with Isolators   92

Hangers Installed through Duct   93

Misaligned Hanger Box   93

Overloaded Spring Hanger   94

Improperly Placed Hangers   95

Overloaded Hanger   96

Vibration-Isolation Hangers   96

Floor-Mounted Pipe Elbow Isolator   97

Incorrect Pipe Penetration   97

Corrected Pipe Penetration   98

Flexible Vibration-Isolation Fitting at Pump   98

Recommended Configurations of Steam Piping for Minimizing
Noise Generation   99

Controlling In-Pipe Noise   102

Controlling Pipe-Radiated Noise   105

Controlling Pipeline Noise with Special Devices   107

Principle of Operation   107

How Pulsatrol Works   107

What Pulsatrol Does   109

Multiple Pipes   **111**

Trapeze Supports for Pipes   111

Trapeze Pipe Isolator   112

Trapeze Pipe Support and Isolation System   113

Pipe Hanger   114

Trapeze Isolators for Large Pipes   114

Pipe Hangers   115

Isolation for Cooling Tower and Condenser Waterlines   116

Multiple-Pipe Penetration   118

BOILERS   **119**

Boiler Isolation   119

AIR COMPRESSORS   **122**

Air Compressor and Tank   122

Air Compressor and Tank   122

Air Compressors   123

COOLING TOWERS AND CHILLERS   **125**

Cooling Tower and Piping without Vibration Isolation   125

Cooling Tower with Spring Isolators   126

Cooling Tower with Spring Isolators   127

Cooling Tower with Spring Isolators of Variable Heights   128

Cooling Tower with Vibration Isolation   129

Cooling Tower with Noise Control Enclosure   130

Cooling Tower with Architectural Screen   131

Inadequate Retrofit for Controlling Cooling Tower Noise   132

Cooling Towers with Silencers   134

Cooling Tower on Concert Hall Roof   135

Cooling Tower Isolation Bridged by Electrical Conduit   136

Steam Absorption Refrigeration Machine   137

Important Details of Installation   138

FANS — SUSPENDED AND FLOOR MOUNTED   139

Fan with Inertia Base and Spring Isolators   139

Resilient Snubbing Device   140

Improper Installation of Hanger-Type Isolator   141

Recommended Methods for Vibration-Isolating Various Other
Fans and Fan Units   142

AIR-HANDLING UNITS   144

Correct Vibration Isolation of Air-Handling Unit   144

Incomplete Installation of Vibration Isolators under
Air-Handling Unit   145

Vibration Isolation of Rooftop Air-Handling Unit   146

Air-Handling Unit with Overloaded Isolation Pads   148

Freestanding Enclosure for Suspended Air-Handling Unit   149

Acoustical Enclosure for Air-Handling Unit   153

Acoustical Treatment of Air-Handling Units   158

Mechanical Equipment Enclosures   160

Exhaust Fan Enclosures   161

UNIT HEATERS   163

Unit Heater Vibration Isolation   163

ROOFTOP EQUIPMENT   164

Rooftop Air-Conditioning Unit   164

Rooftop Equipment on Vibration-Isolation Curb   166

Acoustical Panel Barriers for Rooftop Equipment   168

Rooftop Air-Cooled Condensing Units with Acoustical Panels   170

CONCRETE INERTIA BLOCKS   171

Concrete Inertia Blocks for Floor-Mounted Equipment   171

4. Plumbing Pipes   177

Vertical Plumbing Lines   179

Sanitary Line in Concert Hall Wall   182

Shaft for Ducts and Pipes   183

Vibration-Isolation Soil Lines and Water Pipes   184

Pipe Isolation Hangers for Sanitary Line — 185

Isolation System for Hydraulic Piping in Concert Hall — 185

Resilient Isolation of Plumbing and Heating Distribution Lines — 186

Sanitary Piping with Hubless Fitting and Resilient Pipe Clamps — 187

PLUMBING PUMPS — 190

Vibration Isolation for Plumbing Pumps — 190

Pipeline Vibration Isolators — 192

Sewage Ejector and Sump Pumps — 193

In-Line Pumps — 195

Sump or Ejector Pumps — 196

Sump-Pump Isolation — 197

5. Electrical Conduits and Conductors — 199

Electrical Conduits — 201

Electrical Conductor Penetration — 202

Vibration-Isolating Electrical Conduits — 206

Small-Diameter Flexible Conduit — 207

Large-Diameter ''Flexible'' Conduit — 208

Vibration Test Setup Showing Flexible Electrical Conduit Connector — 209

Electrical Cable Racks — 210

Multiple Electrical Conduits — 211

ELECTRICAL TRANSFORMERS — 213

Small Transformer — Hung — 213

Vibration Isolation for Transformer — 214

Transformer Duct Connection — 216

Transformer Duct Isolators — 217

Portable Transformer Isolation — 218

Fixed Transformer Isolation — 218

Flexible Connections to Transformers — 219

EMERGENCY ELECTRIC GENERATORS — 220

Emergency Electric Generator — 220

Emergency Electric Generator — 221

DIMMERS — 222

Recommended Vibration Isolation for Dimmer Bank — 222

6. Walls, Floors, Ceilings, and Doors — 225

Conventional Methods for Controlling the Transmission of Airborne Excitations — 227

WALLS   228

Resiliently Mounted Wall System   228

Wall Treatment in Mechanical Equipment Room   230

Several Methods of Noise Control Treatment for Walls   231

Structural Separation between Double Walls   234

Utilities in Wall   235

FLOORS   236

Floated Floors   236

Resilient Isolators for Floors   238

Jack-Up Isolators   239

Resilient Structural Separations   240

CEILINGS   242

Duct Connection in Ceiling   242

Resiliently Suspended Ceiling Framing   243

Resiliently Suspended Ceiling with Sway Bracing   243

Vibration Isolation for Ceilings   244

Inertia Hangers for Ceilings   247

DOORS   249

Door Gaskets   249

7. Seismic Isolation and Protection   253

Seismic Protection of Resiliently Mounted Systems   254

Earthquake Force   255

Equipment Mounting for Seismic Protection   257

Seismic Snubber Analysis   259

Seismic Restraining Devices   271

Seismic Isolation for Resiliently Mounted Equipment   275

Automatic Lockout System for Seismic Control   276

Other Design Aids   281

Seismic Isolation and Protection of External Elements   283

Isolation for Gas-Engine-Drive Chiller   284

Snubbers for Pump Isolation System   285

Air Mounts with Seismic Snubbers and Seismic Restraint   286

Correcting a Vibrating Transformer   288

Seismic Restraints for Electrical Transformer   290

Seismic Restraint Isolators with Travel Limit Stops   290

Isolation for Air Compressor   292

Seismic Snubber and Air Mounts   293

Seismic Restraint for Air-Handling Units in Nuclear Plants   294

Turnbuckles and Cable Restraints   295

Seismic Restraints for Hung Piping Systems   296

Seismic Cable Restraints for Hung Piping   299

Seismic Restraints for Piping   302

Isolators for Vane Axial Fan   305

Seismic Snubber Attachment   306

Vibration Isolators with Travel Limit Stops   307

Equipment Isolation with Horizontal Snubbers   308

Field-Fabricated Snubber (Not a Seismic Device)   309

Electric Transformer   310

Position Hanger   310

All-Directional Seismic Snubbers and Snubber Tests   311

Securement for Seismic Restraints   314

8. Special Equipment   317

VIBRATING CONVEYORS   319

Conveyor Systems   319

STAGE LIFT MACHINERY   321

Stage Lift Motor and Drive Unit   321

ELEVATOR MACHINERY   322

Hydraulic Elevator Equipment Isolation, Piping, and Piping
Isolation   322

Inadequate Hydraulic Piping Isolation   324

Motor Generator Set for Elevator   326

Vibration Isolation for Elevator Lifting Equipment   327

Hydraulic Elevator Pumping Unit   328

THEATER AND CONCERT HALL WINCHES   329

Vibration Isolation and Noise Control for Winches   329

BEVERAGE COOLERS   336

Vibration Isolation and Noise Control for Beverage Cooler   336

ESCALATORS   341

Escalator Drive Unit   341

Installation Details and Acoustic Measurements for
an Escalator   344

Decibel Addition   347

APPENDIXES

A Wire Rope Identification and Construction   349

Wire Rope Clips   355

How to Apply U-Bolt Clips   356
Wire Rope Assemblies   363

B   Rivets and Threaded Fasteners   365

C   Smith-Emery Test   375

D   Uniform Building Code — Seismic Excerpts   379

**E   Register 74 — Safety of Construction of
     Hospitals**   388

**F   State of California — Building Standards —
     Health Facilities**   392

**G   State of California — Building Standards —
     Schools**   396

H   SMACNA Standard Details   399

I   Seismic Zone Maps   405

Bibliography   421
Index   423

Foreword

People often think that excessive noise from mechanical equipment in buildings is inevitable. It isn't. Plenty of good hardware for noise and vibration control is available and is specified with good intent. But good hardware and good intent are useless without good installation.

No one knows this better than Bob Jones. In the course of his years of work as a designer of noise control for mechanical systems, he has seen and photographed a wide range of equipment installations, ranging from dreadful to excellent.

This book is for architects, engineers, contractors, building managers, and others responsible for supervising the installation and maintaining of noise and vibration control equipment. It shows how to achieve effective installations and how to correct common errors. These principles and details are illustrated clearly with drawings and pictures from projects all over the world. Careful study of these examples will be of great help to anyone trying to realize the full potential of good design for the control of noise and vibration from mechanical equipment.

ROBERT B. NEWMAN
Bolt Beranek and Newman Inc.

Preface

This book originally started out not as a book but as a job supervision manual to be used on a Far East project by the mechanical engineering consultant at the project. With the illustrated and detailed manual in hand, the local engineer could inspect and supervise the correct installation and application of all noise and vibration control equipment and material for the acoustical consultant in the United States. Any special questions or items that might not be covered by the manual could be handled by express mail, cables, or telexes between the two parties. The use of the manual would make it unnecessary for the U.S.-based acoustical consultant to make frequent visits to the job site.

Once the manual was completed and thoroughly explained during one two-week meeting with the mechanical engineers, it was then used extensively throughout the construction of this multimillion-dollar project.

It was after the presentation of the manual to the engineers that I began to realize more clearly than ever before that there existed a real need to communicate specific instructions to the architect, engineer, and contractors regarding the correct application and installation of all noise and vibration control equipment. A gap existed between presenting the plans and specifications to the contractor and the actual installation of the equipment. It needed to be bridged with written instructions supplemented with detailed drawings, pictures, and catalog cuts showing clearly and completely how to place the equipment and materials correctly so that they would function as designed and so that the recommended acoustical goals would be achieved.

Thus, it was decided to add some additional material, broaden the scope of equipment covered, and turn the results into a book for use not only by architects, engineers, and contractors but also by building developers, building management organizations, private owners, and design-build teams.

I see this book as a means for aiding installers to do the job correctly in the first place. However, if misapplications are found or equipment is installed incorrectly, this book also describes and shows how to correct such situations. In addition, this book assists designers in making accurate selections of good noise and vibration control equipment and materials and explains why and how properly applied equipment works.

ROBERT S. JONES

Acknowledgments

Pat Lama, Mason Industries. Without Pat's extremely valuable experience and practical know-how, the chapter on seismic isolation would not be. The generous gift of his time and expertise is deeply appreciated. Pat is a true and lasting friend. I also want to thank Norm Mason, Mason Industries, for an exceptionally constructive editorial review of this book. Norm's input, particularly with regard to vibration isolation, has enhanced the text in many instances. Deborah L. Melone, Bolt Beranek and Newman Inc. My sincere thanks go to Debbie for her significant contribution of keeping me on-line during development of the book and for organizing material and editorially upgrading the text.

Ewart Wetherill, Peter Terroux, Laymon Miller, Robert B. Newman, Warren Blazier, Carl Rosenberg, Jack Curtis, and Robert Hoover, all colleagues or former colleagues at Bolt Beranek and Newman Inc. Their willingness to teach, share information, and impart real insight and feel for noise and vibration problems and the control of these physical manifestations has been invaluable to me in preparing this book.

Steve Levy, President, and Jim Barger, Chief Scientist, Bolt Beranek and Newman Inc. I appreciate the financial support extended in preparing this book. Even more important to me, I thank them both for their faith in my capability and thus their positive encouragement.

Robert Harvéy, Mary Gillis, and Geoffrey Roever, Bolt Beranek and Newman Inc. Thank you very much for administrative and legal advice and assistance, which you supplied so expertly.

To my secretary, Maureen Fosher, thank you for the many typing tasks that you always managed to work in with your otherwise very busy schedule.

Introduction

1

Noise control consultants generally supply designs and specify equipment for the acoustical treatments they recommend, and equipment manufacturers provide installation instructions with their equipment. However, design specifications and manufacturers' instructions do not always make clear to the installing contractor how to install each treatment correctly. This manual is intended to fill in the gaps in information about noise and vibration control treatments for the contractor, building manager, engineer, or architect unfamiliar with the acoustical treatments used in construction of major buildings. The manual serves as a guide for monitoring installations at all stages of a project.

STRUCTURE OF THE MANUAL

The manual is divided into eight parts. Part 1, Introduction, explains the purpose and structure of the book and defines some basic concepts and terms used in noise and vibration control. Part 2, Guidelines for Architects, Engineers, and Contractors, lists specific requirements for noise and vibration control treatments and describes how they should be installed and maintained. Parts 3 through 8 contain photographs of correct and incorrect installations. These photographs will help readers to determine whether the installations they inspect are correct. Through photographs, drawings, manufacturers' catalog reproductions, and brief descriptions, the manual shows the correct use of acoustical treatments in six major divisions of building construction. With this visual approach, general contractors can see clear examples of typical acoustical problems and their solutions; they can use this information in supervising subcontractors who work on installations for the different systems within a building project. Engineers, architects, and others involved with building construction and management can also use the manual to increase their knowledge of proper acoustical treatments, and of how to maintain such equipment for optimum results.

KEY CONCEPTS USED IN NOISE AND VIBRATION CONTROL

Sound, loudness, pitch, and noise. *Sound* is a form of mechanical energy that moves from a source (a voice, a musical instrument, a machine) along a path to a listener, as tiny oscillations of pressure just above and below atmospheric pressure. When people hear sounds, they can distinguish their *loudness*, their *pitch*, and variations of pitch with time. Some sounds, such as speech, are recognized as having certain meanings and are useful or desirable sounds. Others, such as loud stereo playing when you are trying to sleep or a machine that makes it difficult to hear speech, are annoying or unwanted. Unwanted sounds are called *noise*.

Source, path, and receiver. In treating a noise problem, one must consider the *source* of the noise (a piece of noisy equipment or other noisemaker), the *path* by which the sound travels, whether through air or fluid or through a building structure, and the *receiver*, i.e., the person who hears the noise. For a contractor working on noise control treatments during building construction, there is no way to treat the receiver of the noise. Contractors must concentrate on treating the source of the noise or its path to achieve an appropriate level of sound for all of the future occupants and activities in a building. Because some major building equipment, such as cooling towers, is located outside of buildings and other equipment is located inside, noise problems must be considered both from within the building and from the viewpoint of neighbors outside. The best method of treating the source of a noise, whether it is inside or outside a building, is to specify noise control features and quieted equipment in the original building design. When correcting an already existing noise problem, however, the path of the noise is often the most effective point of treatment.

The descriptions of noise control treatments and installations in this manual contain a number of key terms that are important to understand. Following is a list of definitions of these terms.

KEY TERMS USED IN NOISE AND VIBRATION CONTROL

db: Abbreviation for decibel.

dB(A): A composite abbreviation for decibel (dB) with the indication A to show that an A-weighted filter has been used in the sound level meter to correct for the sensitivity of the human ear. The A-weighted level discriminates against some low-frequency noise.

Decibel: The unit used to express the magnitude, or loudness, of sound. A more precise mathematical definition of a decibel is a unit of level used for the logarithm of the ratio between a variable quantity and a reference quantity.

Dynamic insertion loss (DIL): Acoustical performance with air flowing through a silencer. Sound for determining DIL may be provided by a loudspeaker. Noise generated by air flowing through a silencer is called *self-noise*.

Frequency: Subjectively, frequency is what we associate with the pitch, or tone, of a sound. Frequency is the number of cycles of pressure variations per unit of time and is given in hertz (Hz), or cycles per second (cps).

Hertz (Hz): The internationally accepted unit to express frequency. One hertz is equal to one cycle per second.

Noise criteria (NC) curves: A series of curves based on *criteria* statements of the desired noise or vibration environment. NC curves show the sound pressure levels that should not be exceeded in various human environments. For example, a certain NC curve is recommended for school rooms, another for concert halls, and so forth. NC curves are given in octave frequency bands; they show the maximum permissible sound pressure level for each frequency band of sound (see Figure 1.1).

Self-noise: See dynamic insertion loss.

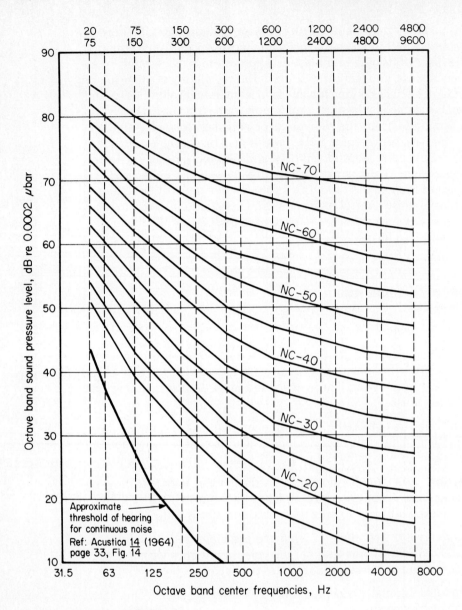

FIG. 1.1 Noise criteria (NC) curves.

Sound power: The total sound energy radiated by the source per unit of time.

Sound power level (PWL): PWL, in decibels, is 10 times the logarithm to the base 10 of the ratio of a given sound power to a reference sound power (standard reference is 10^{-12} W).

$$PWL = 10 \log \left(\frac{W \text{ (watts)}}{10^{-12} \text{ W (watts)}} \right) \text{ dB}$$

Sound pressure level (SPL): A term used to express the magnitude of a sound wave. When the word "level" is used, it means that the quantity being measured is expressed in decibels.

Spectrum: The range of frequencies making up a sound. More specifically, it is the distribution of noise with frequency.

Transmission loss (TL): The sound transmission loss (TL) of a structure is a measure of sound insulation. The TL is related to the logarithm of the ratio of the sound power that passes through a structural element (transmitted sound power) to the total sound power that strikes the element (incident sound power).

Guidelines for Architects, Engineers, and Contractors: Vibration Isolation and Noise Control Treatments

2

GENERAL INFORMATION

Contractors providing vibration isolation and noise control treatments for major buildings must furnish and install vibration and noise control devices, materials, and related items. They are expected to perform work as shown on the drawings and described in the specifications for the job.

Quality Assurance

Contractors are responsible for providing vibration isolators of the appropriate size and capacity to meet the specified deflection requirements and are required to follow the instructions from the manufacturer or vendor of these products. *Deflection* refers to the amount of compression specified for an isolator. For example, if a spring has a minimum static deflection requirement of 1½ in, it should compress at least 1½ in when the equipment load is placed on it and still have adequate clearances between coils. Contractors should consult schedules of equipment on heating, ventilating, and air-conditioning (HVAC) and electrical drawings for the type and specified static deflection requirements of vibration isolators and for the type of mounting required for each item of resiliently supported equipment.

Submittals

For each job, the contractor should submit to the client product data, including manufacturers' specifications and installation instructions, for the following equipment:

- Vibration isolators (include minimum static deflection under various design loads for each type)
- Duct silencers (show certified minimum dynamic insertion loss, self-noise, and pressure drop ratings for each silencer)
- Acoustical duct lining
- Acoustical filler
- Acoustical sealants
- Certified sound pressure or sound power levels for grilles, registers, and diffusers (include the airflow and pressure drop conditions as indicated in the drawings and in accordance with the procedure of ASHRAE*36, latest edition).

The vendor of the vibration isolators is responsible for providing the required project record documents. All such services in connection with the project will be included in the cost of the work.

For each job, the contractor should submit or provide access to samples of the following equipment to the parties the client designates: vibration isolators (each type), acoustical duct lining, acoustical filler, and acoustical sealants. The samples of vibration isolators will be returned for installation at the job after they are approved by the architect or parties designated by the client.

---

* ASHRAE Standard, *Methods of Testing for Sound Rating, Heating, Refrigerating, and Air-Conditioning Equipment*, ASHRAE, New York.

The contractor should also submit to the client shop drawings of the following:

• Location and type of all vibration isolators, with a complete tabulation for each isolator, showing (1) the design load and (2) the minimum static deflection expected under the design load
• Concrete equipment bases and inertia blocks, showing steel reinforcing, steel work, vibration isolators, and anchors
• Location and dimensions of all duct silencers
• Special details at larger scale and all other necessary information to complete the installation.

When the installation is completed, but before substantial completion of the job, the contractor should submit to the client the following project record documents as prepared by an independent consultant:

• A complete tabulation for each vibration isolator, showing (1) the minimum static deflection expected under the design load and (2) the actual static deflection measured at the project
• A report certifying that each piece of operative mechanical equipment does not exceed the specified vibration displacement level

PRODUCTS

Vibration isolators should be produced by a reputable, established manufacturer. A single, approved manufacturer should supply all vibration isolators. Within a stated, reasonable time period such as 15 days after being awarded a contract, and before ordering any materials, the contractor should submit to the architect a complete list of all products proposed for use, including the manufacturer's name and all related product data. Soon after delivery of vibration isolators to the project site, the manufacturer's representative should visit the site and instruct the contractor in the proper installation procedures.

The following list of noise and vibration-isolation equipment describes in detail each item to use.

Floor Isolators

■ **Spring and neoprene pad isolators in series.** These units should be freestanding and laterally stable, without any housing. All mounts should have leveling bolts. The spring diameter should be not less than eight-tenths of the compressed height of the spring at rated load, and springs should be capable of traveling another 50 percent, as a minimum, before becoming fully compressed. This capability is called *additional travel to solid*. This term will be used throughout this section.

Springs should be designed so that the ratio of horizontal stiffness to vertical stiffness ranges between 0.75 and 1.25. Mount each isolator on a double layer of ⅝-in-thick ribbed or waffle-pattern neoprene separated by a 16-gauge galvanized steel or aluminum plate. If necessary, provide a square bearing plate to load the pad uniformly so as not to exceed the manufacturers' limitations and deflect the pads no more than 15 percent of their thickness.

- **Spring and neoprene isolators with travel limit stops.** These isolators should have open, stable, spring mountings, as described in the previous paragraph, and side supports that have vertical travel limit stops to control extension of the spring when the weight is removed or to provide wind resistance in outdoor locations. Travel limit stops are usually, but not necessarily, vertical bolts welded and extended downward from a top plate on each side of the isolators. The bolts pass through large clearance holes on the top of the side plates of the unit. Each bolt acts as a horizontal stop element or snubber and has a supplementary washer on the bottom to prevent upward motion. The housing of the spring units serves as blocking during the erection of the equipment and prevents downward travel. When travel limit isolators must be bolted down or welded in position, it is mandatory that a standard acoustical isolation pad is located between the bottom plate of the spring base plate itself and the base plate of the isolator as welding or bolting would acoustically bypass an external pad without cumbersome precautions. If the unit is not secured to the supports, external cemented pads using two layers of 5/16-in-thick ribbed or waffled pads separated by a 16-gauge galvanized or aluminum plate may be used in lieu of the internal pad.

- **Double-deflection neoprene mountings with straight line deflection curves.** These units require a minimum static deflection of 0.35 in. Cover all metal surfaces with neoprene. Use ribbed neoprene on the top and the bottom surfaces, with bolt holes in the base plates and a tapped hole on top for securing equipment. On equipment such as small vent sets and close-coupled pumps, use steel rails above the mountings to compensate for the overhang when there is one.

Hanger Isolators

- **Hangers for suspended equipment.** Springs in series with neoprene pad hangers should contain a steel spring that is set in a neoprene cup manufactured with a projecting bushing that passes through the hole in the hanger box or is otherwise isolated from the hanger housing to prevent short-circuiting of the hanger rod against the hanger box. The neoprene cup should have a steel washer designed to distribute the load properly on the neoprene and prevent its extrusion. Spring diameters and the lower holes of the hanger boxes should be large enough to permit the hanger rod to swing through a 30° arc before touching the edge of the hole, because such contact between the hanger rod and the edge of the hole will make the spring ineffective. Springs should have a minimum additional travel to solid equal to 50 percent of the rated deflection and diameters equal to no less than eight-tenths of the compressed height of the spring.

- **Springs in series with double-deflection neoprene elements.** These hangers are equal in all respects to the hangers in the previous paragraph, but in addition, they should contain a double-deflection neoprene or fiberglass element in series with the spring, but they should be located against the upper surface of the hanger box. The deflection of this material should be a minimum of $\frac{3}{10}$ in. The addition of the neoprene or fiberglass element is to improve acoustical performance.

▪ **Double-deflection, straight line deflection, neoprene or fiberglass hangers.** These pads should be of 40- to 50-durometer neoprene and sized so that deflections do not exceed 15 percent of the pad's thickness. Standard pads are usually $\frac{5}{16}$ in thick. Solid pads may be used if specifically designed for deflections that do not exceed 15 percent of the pad's thickness. Solid pads should not exceed 60 durometer.

▪ **Waffle-pattern or ribbed neoprene pads.** These pads should be made of 40- to 50-durometer neoprene.

Wall Sleeves

▪ **Metal sleeves containing felt, sponge rubber, or fiberglass.** These acoustical seals should consist of formed and stiffened galvanized steel sleeves or pipe, lined on the inside with $\frac{1}{2}$- to $\frac{3}{4}$-in-thick neoprene, sponge, felt, or fiberglass bonded to the metal sleeves. The inside diameter of the isolation material should equal the outside diameter of the encased pipe in each application; use sleeve lengths equal to the wall thickness or as recommended by the manufacturer for the given diameters, but not less than 2 in long.

▪ **Resilient pipe supports.** These units should be the standard product of the vibration-isolation equipment manufacturer, specifically designed for resiliently supporting vertical pipe risers. Isolation materials should not be less than $\frac{1}{2}$ in thick, and all directional designs should minimize field problems. Materials are generally neoprene with maximum loadings of 500 psi.

Equipment Bases

▪ **Steel frames employing spring isolators for floor-mounted equipment.** Use frames made of structural steel wide-flange (WF) sections sized, spaced, and connected to form a rigid base that will not twist, rack (stretch), deflect, or in any way negatively affect the operation of the supported equipment or the performance of the vibration-isolation mounts. Use frames of adequate size and design to support basic equipment units and motors, as well as any components requiring resilient support to prevent vibration transfer from equipment to the building structure. Such components include:

- Pipe or duct elbow supports
- Electrical control elements
- Other related components

Frames should include side mounting brackets or corner pockets for attachment of the spring isolators. In general, frame depth should be one-tenth the longest dimension of the base but need not exceed 14 in unless structurally requirements dictate deeper sections. The clearance between the underside of any frame or mounted equipment unit and the slab or housekeeping pad below should be at least 2 in.

■ **Steel frame for ceiling-mounted equipment.** These frames should meet the requirements listed in the previous paragraphs and should have provisions for suspensions by steel rods. Isolation is provided by hangers as described in previous paragraphs.

■ **Concrete inertia block framing for floor-mounted equipment.** Concrete inertia blocks should be formed of stone-aggregate concrete (150 lb/ft³) cast within a reinforced steel frame made by bolting or welding formed or structural channels or beams. Inertia block thickness should be not less than one-twelfth the longest dimension of the mounted equipment or equipment assembly. Build the inertia blocks to form a rigid base that will not twist, rack, deflect, crack, or in any way negatively affect the operation of the supported equipment or the performance of the vibration-isolation mounts. Use inertia blocks of adequate size and design to support basic equipment units and motors, as well as any associated components requiring resilient support to prevent vibration transfer from equipment to the building structure. Inertia blocks should have side mounting brackets or pockets to lower the center of gravity and facilitate attachment to spring mountings. The clearance between the underside of any inertia block and the building structure below should be at least 2 in. The vibration isolator supplier should furnish the steel perimeter frame and spring mounting brackets and usually builds in the reinforcement and furnishes anchor bolt templates.

Vibration-Isolation Curbs

Place curb-mounted rooftop air-handling units on a prefabricated, extruded, aluminum-frame vibration-isolation system. Spring isolators should be cadmium-plated, with a minimum of ¾ in static deflection and with 50 percent additional travel to solid. Spring diameters should be no less than eight-tenths of the spring height. Neoprene snubber pads should be located at the corners of the isolator base frame and maintain a minimum clearance of ¼ in so that no restriction to the isolators will occur except under high-wind load conditions. The isolation system should be made weathertight on both the top and the bottom by means of a continuous resilient closure element.

Exterior Metal Parts

All metal parts of vibration-isolation units installed outdoors should be cold- or hot-dip galvanized after fabrication.

Galvanizing should comply with ASTM A 123, A 153, and A 386, as applicable. Cadmium plating is an acceptable alternate.

At the time of shipment to the jobsite, the equipment supplier should submit to the contractor a certified statement by the galvanizer indicating that galvanizing meets the ASTM specifications.

Equipment Mountings

Unless otherwise shown or specified, set all floor-mounted equipment on 4-in-high concrete housekeeping pads. Housekeeping pads are raised concrete blocks placed under equipment to prevent water from rusting metal parts when mechanical equipment rooms are washed down. Housekeeping pads also prevent rust if corrosive fluids leak or spill from containers or piping systems. Mount vibration isolators and concrete inertia blocks on the concrete pads.

No equipment unit should bear directly on vibration isolators unless its own frame is rigid enough to bridge across the isolators and unless such direct support is approved by the equipment manufacturer. All support frames should be rigid enough to prevent incorrect alignment of components installed on them.

Unless otherwise indicated, all equipment mounted on vibration-isolated bases should have a minimum operating clearance of 1 in between the inertia base or structural steel frame and the concrete housekeeping pad or floor beneath the equipment. The contractor should check the clearance space to ensure that no construction debris has been left to prevent or restrict the proper operation of the vibration-isolation system.

Mechanical equipment with operating weight substantially different from installed weight (such as refrigeration machines, boilers, and cooling towers) and equipment subject to wind loads (such as cooling towers, large condensing units, and other rooftop equipment) should be mounted on spring isolators with travel limit stops. Cooling tower and condensing unit isolators should be installed to imitate the pattern of support provided for a rigidly mounted tower. In most cases, steel beams are needed between the equipment and the unit isolators, so mounts are usually located under this beam or grillage. Smaller units may be directly mounted on the isolators if agreeable to the manufacturer. The installed and operating heights of the vibration-isolation equipment should be identical. Maintain a minimum clearance of ¼ in around restraining bolts and between the housing and the spring to prevent interference with the spring action. Keep travel limit stops out of contact with isolator top plate and bolts during normal operation.

Isolate miscellaneous pieces of mechanical equipment (such as domestic hot-water converters, heating converters, storage tanks, condensate receiver tanks, and expansion tanks) from vibration on neoprene isolators with a minimum static deflection of 0.35 in, but only when specifically noted on the drawings.

Duct Silencers

Duct silencers should be prefabricated standard products of a responsible manufacturer.

Make the outer casing of rectangular silencers of 22-gauge galvanized steel in accordance with the ASHRAE Guide's recommended construction for high-pressure rectangular ductwork. The seams should be lock-formed and filled with mastic. Make the interior casings for rectangular silencers of not less than 26-gauge galvanized perforated steel.

For filler, use inorganic mineral or glass fiber dense enough to obtain the specified acoustic performance and be packed under not less than 5 percent compression. Material should be inert, verminproof, and moistureproof.

Use the following combustion ratings for filler when tested in accordance with ASTM* E 84, NFPA* standard 255, or UL* no. 723:

Minimum Ratings	
Flame spread	<25
Smoke development	<50
Fuel contribution	<50

When required, provide additional airtightness by applying duct sealing compound at the jobsite. Silencers should not fail structurally when subjected to a differential air pressure of 8-in water gauge from the inside to the outside of the casing.

Determine the acoustic rating of silencers in a duct-to-reverberant-room test facility that provides for airflow in both directions through the test silencer during rating. The test setup and procedure should eliminate all effects caused by end reflection, directivity, flanking transmission, standing waves, and test chamber sound absorption. Acoustic ratings should include dynamic insertion loss (DIL) and self-noise (SN) power levels with airflow of at least 2000 ft/min entering face velocity.

Static pressure loss of silencers should not exceed the requirements listed in the silencer schedule on the drawing at the airflow indicated. Make airflow measurements in accordance with applicable sections of ASME*, AMCA*, and ADC*, for airflow test codes.

Acoustical Duct Lining

Do not install any duct lining damaged in shipment. Acoustical duct lining should not impart any odor to the air. It should not delaminate (separate into layers) or be loosened by the airstream under normal operating conditions.

Use the following duct liner combustion rating, in accordance with ASTM E 84, NFPA standard 255, or UL no. 723:

Minimum Ratings	
Flame spread	<25
Smoke development	<50
Fuel contribution	<50

---

* ASTM	=	American Society for Testing and Materials.
NFPA	=	National Fire Protection Association.
UL	=	Underwriters Laboratories Inc.
ASME	=	American Society of Mechanical Engineers.
AMCA	=	Air Moving Contractors Association.
ADC	=	Air Diffusion Council.

Acoustical Sealant

Concealed application. Use nonshrinking, nonmigrating, nonstraining sealant, either nondrying or permanently elastic.

Exposed application. Use permanently elastic paintable acoustical sealant, either latex, acrylic, or acrylic-latex.

EXECUTION

Installation of Vibration-Isolation Devices

Select locations of all vibration-isolation devices for ease of inspection and adjustment as well as for proper operation.

Place vibration-isolation hangers so that hanger housing may rotate a full 360° without touching any object.

Resilient Support of Piping

Provide resilient support for ceiling-hung piping as follows:

- At the first three support points of pipe runs connected to vibrating equipment, use spring and neoprene element isolators with the same deflection as specified for the equipment. Deflection of the remaining pipe supports shall be a minimum of ¾ in and shall continue throughout the structure.
- Provide resilient support for floor-supported piping at the same locations as specified for ceiling-hung piping. When ceiling support is not practical, use combination spring and neoprene pad isolators for the first three support points with the same deflection as the isolated equipment and with a minimum of ¾-in deflection thereafter.

Resiliently isolated vertical pipe runs. Attention must be paid to movements caused by expansion and contraction. In order to prevent large shifts of loads, isolator deflections should be a minimum of 4 times the anticipated movement at support points. Where piping is guided between such resilient supports, all directional resilient pipe guides should be used. Where actual anchors must be used, all directional resilient anchors should be attached to welded clamps or other devices that are rigidly attached to the piping.

CAUTION: *Under no conditions should piping, ductwork, or conduit be suspended from one another or touch one another. Vibrating systems should be kept separate from nonvibrating systems.*

Connections of Piping

Piping connected to vibration-isolated equipment should be installed so that it does not strain or force vibration isolators supporting the basic equipment out of alignment. Pipes should not restrict such equipment from floating freely on its vibration-isolation system.

Drainpiping connected to vibrating equipment should not touch any building construction or any systems or components that are not vibration-isolated.

Resilient Support of Ducts

Provide resilient support for ceiling-hung ducts where it is specifically called for on the drawings or specifications and to the extent described. Ducts should be resiliently isolated by either spring and pad hangers for hung ducts or springs and neoprene pad isolators for floor-supported ducts.

Connections of Ducts

Connect ducts to fan intakes and discharges by means of flexible connectors made of neoprene, leaded vinyl, or canvas. The face length of the flexible connector should be approximately 4 in. Flexible material should not be so slack that it becomes concave during fan operation, thus interfering with airflow.

Operation of Rotating Equipment

Fans and other rotating mechanical equipment should not operate at speeds in excess of 80 percent of their critical speed.

Balance rotating equipment, such as fans, according to the following schedule for vibration displacement. For actual installed conditions, measure the vibration levels with the equipment installed on its vibration isolators.

Vibration Displacement, Mils Peak to Peak

Speed, r/min	Preferred	Acceptable
Under 600	2	4
600 to 1000	1.5	3
1000 to 2000	1	2
Over 2000	0.75	1

Installation of Duct Silencers

Install a flexible semirigid duct liner where shown by cross-hatching or other designation on the drawings. Install duct liner in accordance with the Sheet Metal and Air-Conditioning Contractors National Association's Duct Liner Application Standard (latest edition).

Penetrations of a Structure

When pipes and conduits with diameters over 12 in as well as ducts penetrate a structure, tightly pack the space between the penetrating member and the building construction with fibrous material the full depth of the penetration and seal both sides of the penetration with a nonhardening, resilient sealer. The space around penetrating members should be limited to between ½ and 1 in. If spaces left are larger than 1 in, close in the additional space to between ½ and 1 in, using materials that provide at least the same surface weight as the structure being penetrated. Pipes and conduits with diameters through 10 in shall be isolated where they penetrate a structure by installing factory-fabricated metal sleeve devices as described under wall sleeves.

Mechanical Systems: Heating, Ventilating, Air Conditioning (HVAC)

VIBRATION ISOLATORS

HANGERS

3.1 Vibration-Isolation Hanger _____

3.2 Figure 3.1 is a correct type of vibration-isolation hanger
3.3 properly installed except for one detail. The electrical
conduits should not be so close that they touch the
hanger box as it may rotate on its top hanger rod due to
equipment vibration. When installing vibration-isolation
hangers, leave enough space around the hanger so that
the hanger box can rotate or turn freely a full 360°
horizontally. These boxes do move as the equipment
vibrates.

The spacing between the coils of the spring indicates
clearly that it is not overloaded, and the large square
hole in the bottom of the hanger box allows excellent
clearance (at least a 30° total arc) before the bottom rod
can touch the hanger box (see Figure 3.2 on page 22).

Applying vibration-isolation hangers so that the hanger
box extends up into the legs of a structural steel channel,
as seen in Figure 3.3, is not good practice.

If there is an upper neoprene element in series with
the spring, as in the illustrated case, it may be next to
impossible to inspect the element for performance.

As in other instances, previously noted, the hanger box
is not free to turn on its hanger rod without contacting
the legs of the channel iron. Once such contact is made,
the isolator can become, at least in part, bypassed or
short-circuited with respect to isolation capability.

FIG. 3.1   Vibration-isolation hanger.

FIG. 3.3   Vibration-isolation hanger installed in
a channel-iron support.

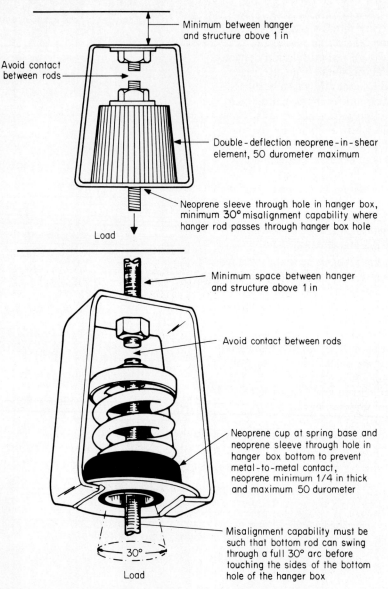

Minimum between hanger
and structure above 1 in

Avoid contact
between rods

Double-deflection neoprene-in-shear
element, 50 durometer maximum

Neoprene sleeve through hole in hanger box,
minimum 30° misalignment capability where
hanger rod passes through hanger box hole

Load

Minimum space between hanger
and structure above 1 in

Avoid contact between rods

Neoprene cup at spring base and
neoprene sleeve through hole in
hanger box bottom to prevent
metal-to-metal contact,
neoprene minimum 1/4 in thick
and maximum 50 durometer

Misalignment capability must be
such that bottom rod can swing
through a full 30° arc before
touching the sides of the bottom
hole of the hanger box

30°

Load

FIG. 3.2 **Hanger mounts.** *(a)* **Neoprene-in-shear hanger.** *(b)* **Spring hanger.**

3.4 Vibration-Isolation Unit

3.5 The spring vibration-isolation unit pictured in Figure 3.4 is a correct type. However, the spring is overloaded, compressing the coils so much that a piece of paper cannot fit between them. The neoprene elements both under and above the spring are also overcompressed from the excessive load imposed on them. Note also that the hanger box cannot rotate horizontally through the full 360° recommended without touching the concrete structure near the top of the box. All of these installation problems contribute to making the unit ineffective (see Figure 3.5).

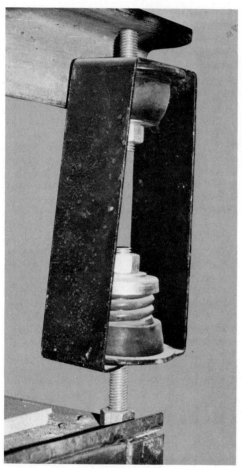

FIG. 3.4   Vibration-isolation unit.

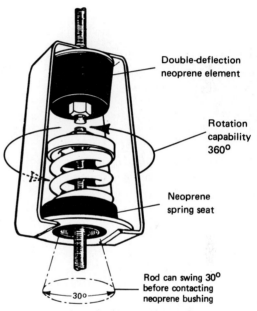

Double-deflection neoprene element

Rotation capability 360°

Neoprene spring seat

Rod can swing 30° before contacting neoprene bushing

FIG. 3.5   Steel spring in series with neoprene hanger: hanger box rotation 360°. *(Mason Industries.)*

3.6 Vibration-Isolation Hanger for Air-Handling Unit

Proper alignment and exceptionally good clearance between hanger rod and isolator box are the important items to observe in this installation (Figure 3.6). Notice, too, that the hanger box is placed so that it may rotate a full 360° horizontally without touching nearby utilities or the building structure.

Easy visual inspection and open access to check the spring static deflection are provided by this carefully arranged installation.

FIG. 3.6   Vibration-isolation hanger for air-handling unit.

3.7 **Vibration-Isolation Hangers**

3.8 The hanger shown in Figure 3.7 is well-designed. Its
3.9 bottom rod can swing through a full 30° arc before
touching the sides of the hole at the bottom of the hanger
box. The steel spring in this isolator sits in a specially
designed neoprene cup, which is made with an extended
lower collar to provide a resilient sleeve through the
bottom opening of the steel hanger box.

However, this unit was so badly misaligned during
installation that the sharp threads of the lower rod have
already cut through the neoprene protective collar.
Without competent job supervision, errors like this one
occur no matter how well a vibration-isolation unit is
designed and manufactured.

Vibration-isolation devices are often badly overloaded.
This is the case in Figure 3.8, where both isolating
elements—the steel spring and the neoprene element—
are compressed to the point that they are totally ineffective. Note the bulging around the edge of the neoprene
and the tightly compressed coils of the spring in this
picture.

In Figure 3.9, the isolating unit is even more overloaded
than in Figure 3.8, so badly that the neoprene element at
the top of the hanger has fractured and come apart. The
excessive weight placed on this unit has visibly bowed
the top and bottom sections of the hanger box frame.

FIG. 3.7 **Badly misaligned hanger.**

FIG. 3.8 **Badly overloaded hanger.**

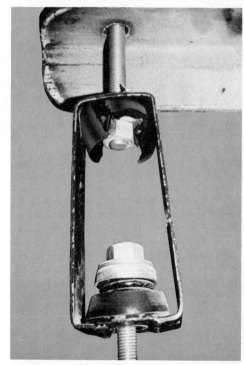

FIG. 3.9 **Hanger with fractured neoprene element and totally compressed spring.**

3.10 Vibration-Isolation Hangers _____

3.11
3.12
3.13

Many manufacturers of vibration-isolating equipment are now producing hanger units with lower rods that can move through a full 30° arc before touching the bottom hole in the hanger box (Figure 3.10). This design feature seems practical and sensible. Although this feature is intended to make the units more foolproof, it does not prevent incorrect installations. In this picture, the bottom hanger rod has been forced against the side of the bottom hole in the hanger box. Therefore, the spring isolator cannot provide vibration isolation because the metal parts of the isolator are touching. When considering this type of hanger, take care to select the springs for maximum stability.

A neoprene bushing extending into the hole in the bottom of this hanger would have helped to minimize vibration transmission.

Figure 3.11 shows a fan in a field house in Halifax, Nova Scotia. At the left side of the picture is the exterior housing of an axial fan to the end of which a square, steel-framed support plate has been welded. Note especially the lower threaded rod coming from the vibration-isolation hanger. This rod extends so far that it is forced against the fan housing (see arrow), pushing the spring of the isolator out of alignment. The spring's coils are thus pressed tightly together, making the unit ineffective as a vibration isolator. Fortunately, the remedy is quite easy in this instance, even though there are four hangers supporting the fan. The four hanger rods can easily be cut off just below the flange of the fan support plate, freeing the isolators to operate properly.

Figure 3.12 shows one of several vibration-isolation hangers that had already been inspected by the architect's representative on a project. The spring and neoprene collar, the essential elements of the unit, are dangling uselessly from the bottom of the hanger box. Members of the design team who inspected this installation had judged the job finished and "punch-listed" it complete with regard to installation. The acoustical consultant included on the design team had no authority to inspect the installation, in this case.

These faulty installations were discovered only because the acoustical consultant retained to make recommendations for noise and vibration control visited the job at personal expense. A qualified acoustical consultant authorized to inspect the job could have discovered and aided in correcting such errors in just an hour or two of on-site inspection.

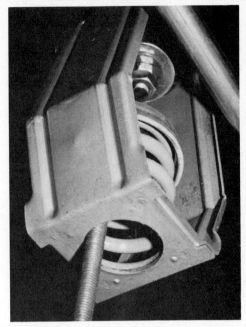

FIG. 3.10   Vibration-isolation hanger.

FIG. 3.11   Isolation hanger for axial fan.

Vibration-isolation hangers that rest against the vibrating equipment they are supposed to isolate cannot work effectively. Figure 3.13 illustrates this problem. When installing such a device, it is good practice to place the unit so that the hanger box can rotate horizontally a full 360° without touching either the equipment being isolated, other equipment or utility systems, or the building structure. In this case, the hanger box rests directly against the housing of this axial fan.

The problem could have been avoided by raising or lowering the hanger box in the existing hanger rod.

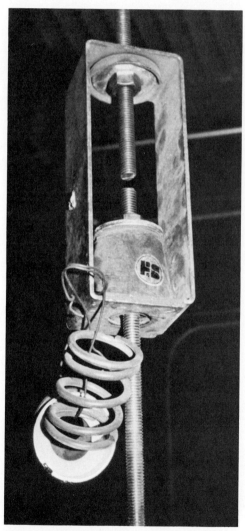

FIG. 3.12   "Inspected" hanger unit.

FIG. 3.13   Hanger: misapplied installation.

HOUSED SPRING ISOLATORS

3.14 Housed Vibration Isolator _____

3.15 Figure 3.14 shows a housed isolator under a fan in a
laboratory building in Louisiana. Housed or semihoused
vibration isolators are not recommended because the
equipment malfunction shown in this picture is very
common. These units are designed so that the inner
guide members, which are an integral part of the top
plate, will ride freely inside each of the two channel-
shaped bottom housings (see Figure 3.15). Neoprene pads
are affixed to the inside of the housings to prevent
contact between metal parts when the usually unstable
springs tip or become misaligned. In fact, most of the
springs do tip, and the weight of the equipment often
causes the metal parts of the isolator to cut through the
neoprene pads, making the spring mounts ineffective. In
Figure 3.14, the top plate of the isolator directly touches
the side housing because of the unstable spring (see
arrow), allowing severe misalignment and making the
isolator ineffective.

Open, stable, steel spring isolators, with snubbing or
lateral movement restrictors when required, should be
used to prevent the problems shown here. These isolators
may be used on the equipment base (see Figure 3.33) or
may be used across the flexible duct connections to fans
and air-handling units (see Figure 3.42).

FIG. 3.14   **Housed vibration isolator.**

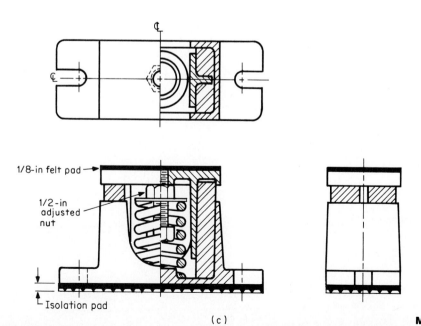

Operating height

Section

(a)

FIG. 3.15 **Housed vibration isolators.** *(Korfund Dynamics Corporation.)*

Ribbed neoprene pad

Section

(b)

⊄

⊄

1/8-in felt pad

1/2-in adjusted nut

Isolation pad

(c)

3.16 Housed Spring Isolators

To perform most effectively, spring isolators must be selected for the proper weight loading. Figure 3.16 is a photograph of two vibration isolators under the same piece of mechanical equipment and shows one spring properly weight-loaded (spring coils open) and the other one overloaded (spring coils tightly compressed). The semihoused isolators shown here are not recommended because they are too difficult to maintain in proper alignment.

The overloading of one spring in the picture with almost no load on the other could be the result of poor adjustment rather than insufficient capacity.

FIG. 3.16   Housed spring isolators.

3.17 Housed Vibration Isolator

The isolator shown in Figure 3.17 is used in a multilevel office building in Quebec. Close examination of the installation revealed that the effectiveness of the spring was reduced because the inner side casting of the top assembly came into direct contact with the outer side casting of the bottom assembly (see arrow). This problem occurs frequently because of misalignment of the top and bottom castings in the field, uneven load distribution, and misalignment caused by starting and stopping of mechanical equipment and instability in the spring element itself. This isolator is not designed to correct for the installation hazards described above.

One reason manufacturers and engineers do not recommend isolators of this type is that the units are difficult to inspect to see if they are functioning properly, as Figure 3.17 shows.

Observation of this type of isolator on many projects has shown that they are often practically useless, because the neoprene guide sleeve, which is supposed to keep the top and bottom castings separate, is either too small or is easily crushed or cut through, allowing contact between top and bottom housings. Sometimes misalignment or tipping of the top casting is so severe that the side guide of the top casting is actually pressed down against the base of the bottom casting, making the spring isolator totally ineffective.

FIG. 3.17   Housed vibration isolator.

3.18 Housed Isolators
3.19 Figures 3.18 and 3.19 are two more examples of the
dangers of using housed isolators. Both pictures show *end*
views of spring isolators, clearly illustrating how the
units are misaligned so that vibrations are transmitted
directly from the top of the unit to the base and from
there to the building structure.

In Figure 3.18, the top casting is tipped within the side
members of the bottom casting so that direct contact is
made (see arrow). In Figure 3.19, the top casting, though
not so badly tipped, has shifted so far to the right that it,
too, rests firmly against one side member of the bottom
casting (see arrow). The U-shaped foam rubber element is
supposed to prevent such metal-to-metal contact, but as
you can see, it does not work that way in the field.

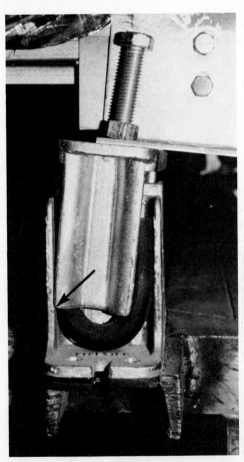

FIG. 3.18 **Housed isolator.**   FIG. 3.19 **Housed isolator.**

3.21
3.22
In the type of isolator shown in Figures 3.20 to 3.22, the cast-steel guide member extending downward invariably comes into direct contact with the two guide members that protrude upward from the isolator base plate. This problem may occur because the units were misaligned during installation, or it may occur later, even if the units are properly aligned when installed, because of the starting, stopping, or normal operating motion of equipment. For example, the housed isolators under an air-moving fan may be perfectly aligned when installed, but when the fan motor is turned on and the fan reacts to pressures imposed by the connected duct system, the fan and its base often move. When this happens, the isolators become tipped or misaligned and are thus ineffective as vibration isolators.

Several manufacturers produce versions of housed spring isolators, each differing slightly from the others. However, regardless of differences in construction, all housed isolators have the same inherent disadvantages. They easily become misaligned, and top and bottom elements come into direct contact with each other, thus destroying the effectiveness of the units. Figures 3.20 to 3.22 are similar except that each isolator serves a

FIG. 3.20 Housed isolator rusting together.

FIG. 3.21 Tipped isolator.

FIG. 3.22 Badly tipped isolator.

different piece of mechanical equipment. These three similar pictures are shown to emphasize how frequently problems with housed isolators occur. The following details of these installations reveal common problems of housed spring isolators.

In Figures 3.20 and 3.21, the top casting (guide member) has become wedged firmly between the castings coming up from the base of the isolator, and the two parts are becoming rusted together (see arrows). This direct contact between the two parts has occurred despite the neoprene sponge separators cemented in place at the factory. These neoprene separators inside the guide members are of little or no value in maintaining resilient separation between the top and bottom elements of the isolator, because they slip out of place or are crushed so that they are no longer useful. In the project where these installations were photographed, about 85 percent of approximately 200 similar units were malfunctioning as shown in these photos.

In Figure 3.22, the unit is so poorly aligned that one of the neoprene separators, supposedly cemented in place, has fallen out (see arrow), and the other one inside the housing has been crushed so much that it cannot provide any separation between the top and bottom parts of the isolator.

Where resilient bolting of such isolators to the floor was needed on this project, none was provided. Generally, it is not necessary to bolt isolators like these into place. When the base plate of the isolator is not bolted down, it is fairly easy to realign the bottom component with the top. But as soon as the equipment being isolated begins to operate, the isolators are easily thrown out of alignment again.

Housed isolators malfunctioning like those in Figures 3.20 to 3.22 show up in installations all over the world, despite all the published claims of isolator manufacturers that such units function well in the field. This is not to say that housed isolators never work, for in some cases they do. However, years of experience in supervising and inspecting projects have convinced the author that the housed isolators are not likely to be installed and maintained correctly, so that they provide the degree of vibration isolation required. This series of pictures focuses on only one disadvantage of housed isolators. Other problems are considered in Figure 3.23.

3.23 Housed Isolator with Multiple Springs

The housed isolator shown in Figure 3.23 differs from those shown earlier because it has several springs. However, the problems that occur with these isolators are typical of single-spring units as well. The springs are overextended here because of improper loading, causing some misalignment. In such situations, when the springs are extended to their maximum, they become laterally unstable and can tip more easily than when compressed. Tipping allows the top and bottom parts of the isolator to touch and form a rigid connection instead of being isolated from each other. If the springs on this unit were extended a little more, the equipment mounted above it could topple over. Figure 3.23 illustrates the importance not only of proper installation but also of the correct selection of each isolator to ensure proper weight-load distribution wherever an isolator is installed. Isolators like these should not be used unless castings overlap by at least 1 in before installation.

FIG. 3.23 Housed isolator with multiple springs.

UNHOUSED SPRING ISOLATORS

3.24 Open Spring Isolator

In contrast with the housed spring design shown in Figure 3.17, Figure 3.24 shows an open, stable, steel spring unit. Open spring units may be inspected easily, and the static deflection of the spring under load may be measured with any ruler.

A hazard not easily noticed on housed isolators but easily seen in open spring units like the one shown here is that the adjustment rod, the threaded bolt extending down through the center of the spring, is sometimes so long and so poorly adjusted that it touches or pounds on the bottom plate. In this installation, the bolt had to be screwed upward until it was far enough from the base plate so that no contact would occur between the two parts of the isolator during normal operation, including starting and stopping of equipment. Sometimes the bolt requires shortening.

This problem may occur in both housed and open spring isolators, but it is much easier to detect in open units and therefore more likely to be corrected during the job supervision stage of a project.

FIG. 3.24 Open spring isolator.

3.25 Single-Spring Vibration-Isolation Unit

3.26 This correctly installed unit uses *travel limit stops*, four bolts located two on each side of the unit. When installed properly, the travel limit stops ride freely through oversized holes in the top plate of the side channels of the unit. The nut and washer of each bolt have been loosened to allow the spring to pick up its load. The nuts and washers act as the travel limit stop elements, also called *snubbers*.

For the most effective operation of the spring and the travel limit stops, the unit must be aligned perfectly. The top plate of the unit must be secured firmly to the vibrating equipment, with at least ½ in clearance between the top of the side channel frames and the top plate. Figure 3.25 clearly shows this recommended clearance between the side channels and the top plate of the unit.

To reduce high-frequency vibrations that may be transmitted through the coiled spring, two ⁵⁄₁₆-in-thick layers of ribbed or waffle-pattern neoprene pads are installed under the entire base plate of the isolation unit. To secure the isolation unit to the steel frame on which it sits, install hold-down bolts on outdoor equipment subject to wind loads. These bolts should be installed as shown in Figure 3.26a, using resilient grommets similar to those shown in Figure 3.26b and c. Better results may be obtained by using an internal pad when the mountings must be secured so that welding or bolting does not short-circuit the conditions.

FIG. 3.25 Isolator with travel limit stops.

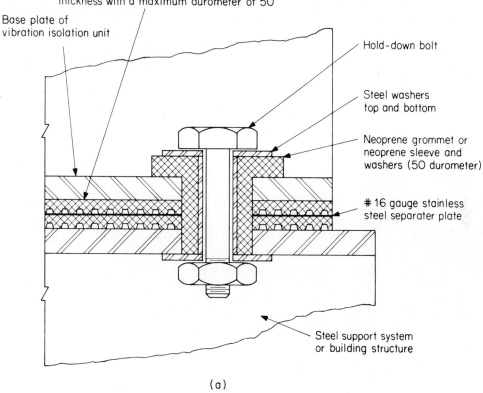

Provide two layers of ribbed or waffle-pattern neoprene pads. Pads should be of 5/16 in minimum thickness with a maximum durometer of 50

Base plate of vibration isolation unit

Hold-down bolt

Steel washers top and bottom

Neoprene grommet or neoprene sleeve and washers (50 durometer)

#16 gauge stainless steel separater plate

Steel support system or building structure

(a)

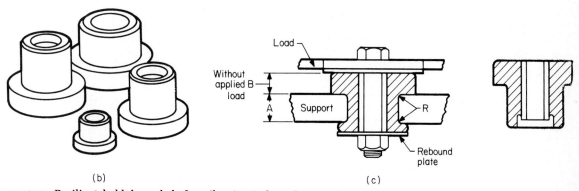

(b)

Load

Without applied load

B

A

Support

R

Rebound plate

(c)

FIG. 3.26 Resilient hold-down bolt for vibration-isolator base and grommet installation (no scale). *(a)* Detail of bolt. *(b)* Neoprene grommets. *(c)* Typical grommet installation. *(Uniroyal.)*

3.27 Double-Spring Vibration Unit

This double-spring isolator (Figure 3.27), similar to the one in Figure 3.25, illustrates that installations may be made correctly at many different geographical locations by competent installers. This picture of an isolator in a high-rise office building in Georgia shows the fine alignment in this installation and the correct clearance between the tops of the side channel frames and the top plate of the unit.

However, concrete grout has spilled over onto the isolator base plate and the springs. The grout has also spilled onto the neoprene pads under the isolation base plate, somewhat reducing their effectiveness. Such spillage should not be allowed to remain on a project. If it occurs, the contractor responsible should be instructed to clean off the grout and any other foreign material before the installation is approved.

FIG. 3.27 Double-spring vibration unit.

3.28 Double-Spring Isolator

The center adjusting rod of the isolator shown in Figure 3.28 was screwed upward to allow the springs to accept the weight above it. Nevertheless, the excessive weight load has completely compressed the spring coils, thus overloading the springs. The travel limit stop bolts and washers at each side of the isolator housing were loosened correctly, and yet the unit cannot provide isolation because of the excessive load on it.

Even when weight selection is correct, if there are solid pipe connections, the contractor does not always provide proper pipe support so that the piping weight causes the overload. In other examples, when flexible connections are used, pressure may force them to elongate forcing the isolators solid.

This problem occurs when equipment point loads are not carefully checked and the isolators are not matched properly with the loads. To prevent this error, spring coils are usually color coded so that the installer can select the right isolator for each weight load. Installing mechanics should make sure they have isolators that can accommodate the actual weight they must carry. Otherwise, the units may have to be replaced, as this one was.

Isolator manufacturers color-coded springs indicate to the installer what maximum weight load the unit can carry. Having a weight-loading diagram for the equipment being isolated, the contractor can easily determine the correct position under the equipment for each color of isolator spring.

FIG. 3.28 Double-spring isolator.

3.29 Spring Isolator for Cooling Tower

In Figure 3.29, the vertical side support of the isolator is touching the channel-iron support of the cooling tower base (see arrow), though the spring is not overloaded and the travel limit stop bolts have been loosened correctly. Problems like this usually occur when springs with travel limits stops are installed in a tipped position or supporting members above or below the isolator are not plumb. Contrast this installation with the correct one shown in Figure 3.27.

The hard-bolting of the base plate of the isolator to the steel dunnage support system is permissible in this instance because the manufacturer of this unit provides an additional neoprene pad between the spring plate and the isolator base. However, the hold-down bolt installation method should still include resilient separation (see Figure 3.26*a*, *b*, and *c*).

FIG. 3.29 Spring isolator for cooling tower.

3.30 Vibration Isolator with Travel Limit Stops

Although the vibration isolator in Figure 3.30 is installed and adjusted correctly, it still cannot work to full advantage until the travel limit stop bolt is aligned in the center of the hole through the top plate of the side support (see arrow). This problem emphasizes the need for correct alignment of all parts of an installation before it is secured.

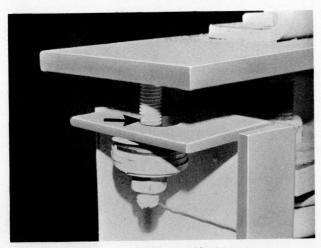

FIG. 3.30 Vibration isolator with travel limit stops.

3.31 Spring Isolator with Unreleased Travel Limit Stops

Figure 3.31 shows an example of a spring isolator in which the travel limit bolting mechanism has not been released. The desired ¼- to ½-in clearance between the travel limit stops and top plate of the isolator is therefore nonexistent (see arrow). In this position, the isolator performs no vibration-isolating function.

FIG. 3.31 Spring isolator with unreleased travel limit stops.

3.32 Spring Isolator for Fan

3.33 The problem with this isolator was not detected until an acoustical consultant stepped on the fan base, adding extra weight to the isolator, and thus found that nothing moved. Getting down on hands and knees and peering around the soil pipe at the right and looking under the fan base, the consultant discovered that the travel limit bolt at the left extended from the base plate of the isolator to the underside of the fan base (see arrow). The bolt completed the path from the vibrating fan to the concrete housekeeping pad, which rests directly on the floor slab of the mechanical room, thus allowing vibrations to travel from the fan to the floor. Although there is a neoprene separator between the isolator base plate and the housekeeping pad, the neoprene cannot provide the high-frequency isolation needed for fan installations.

This type of isolator can be hard-bolted in place as in Figure 3.32, because the spring itself sits in a neoprene cup instead of being welded to the isolator base plate. If there is no neoprene cup, use a resiliently treated bolting arrangement like the one shown in Figure 3.33.

FIG. 3.32 **Spring isolator for fan.**

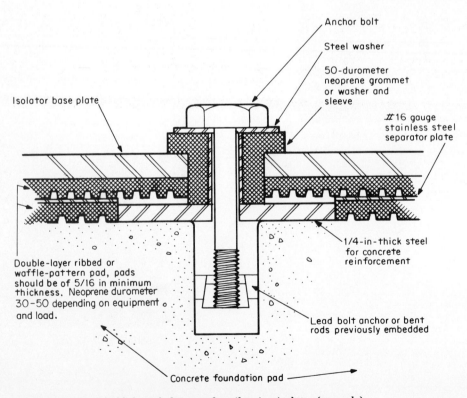

FIG. 3.33 **Resilient hold-down bolt insert for vibration isolator (no scale).**

Anchor bolt

Steel washer

50-durometer neoprene grommet or washer and sleeve

∄16 gauge stainless steel separator plate

Isolator base plate

1/4-in-thick steel for concrete reinforcement

Double-layer ribbed or waffle-pattern pad, pads should be of 5/16 in minimum thickness. Neoprene durometer 30-50 depending on equipment and load.

Lead bolt anchor or bent rods previously embedded

Concrete foundation pad

3.34 Spring Isolator with Lateral Restraint

Figure 3.34 shows a combination spring isolator with a built-in neoprene element for restricting lateral movement of the fan unit mounted above it. To limit lateral movement of point-isolated equipment and when using devices like the one shown here, resiliently bolt the isolator in place.

In such cases, hold-down bolts should be resiliently treated as shown in Figure 3.33.

FIG. 3.34 Spring isolator with lateral restraint.

3.35 Misaligned Spring Isolators
3.36

Installing spring isolators of any type so that the natural alignment of the spring coils is correct from top to bottom is essential if the spring is to provide its rated static deflection under load. Figures 3.35 and 3.36 show how poor alignment of the springs prevents them from working properly. When a spring isolator is forced out of alignment or tipped from a true vertical centerline, the spring coils tend to come closer to each other until they touch (see arrows). At this point, the spring is virtually useless in providing vibration isolation.

Another problem caused by faulty alignment occurs with the travel limit stop bolt. In these examples, the bolt extends from the base plate of the isolator up through the top plate, where it is capped with a nut. If the unit is aligned correctly, the top channel of the spring moves freely. However, the movement may cause the bolt to touch the side of the opening in the top channel through which it passes. Unwanted vibrations can travel along this path, making the spring isolator ineffective.

FIG. 3.35 Springs misaligned.

FIG. 3.36 Springs badly tipped.

3.37 Unhoused Spring Isolator

3.38
3.39 Earlier figures in this section have shown the problems of using housed spring isolators. Figure 3.37 shows that problems can occur even with an unhoused isolator if it is not properly constructed or installed.

The isolator shown in Figure 3.37 is in a hosptial in Detroit, where many types of vibration-sensitive equipment are used. The construction details highlight the need to pay attention to details when fabricating isolator supports. In this installation, a cantilevered top support bracket was welded to rectangular rather than angle-cut side plates. When the unit was installed, the side plates rested firmly on the base plate on both sides, forming a partial housing and touching the base plate, so that no vibration isolation could occur.

It is most practical for installations like this to use cantilevered brackets, wider brackets, or brackets requiring no gussets.

To allow the isolator to work properly, the contractor burned off the corners of the side plates so that they would not touch the base plate. However, this minimal corrective work was inadequate. If a person merely

FIG. 3.37 **Unhoused spring isolator.**

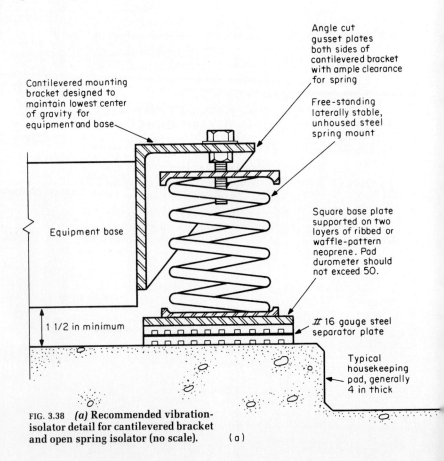

Cantilevered mounting bracket designed to maintain lowest center of gravity for equipment and base

Equipment base

1 1/2 in minimum

Angle cut gusset plates both sides of cantilevered bracket with ample clearance for spring

Free-standing laterally stable, unhoused steel spring mount

Square base plate supported on two layers of ribbed or waffle-pattern neoprene. Pad durometer should not exceed 50.

⌗ 16 gauge steel separator plate

Typical housekeeping pad, generally 4 in thick

FIG. 3.38 *(a)* **Recommended vibration-isolator detail for cantilevered bracket and open spring isolator (no scale).** (a)

placed a foot on the equipment base, it could cause the side plates to touch the base plate again and stop the isolator from working. This example emphasizes the need to comply with the detail shown in Figure 3.38a and provide angle-cut gusset plates for the sides of cantilevered brackets.

This example should illustrate that it is better to do a job right the first time than to do it half right more than once and still not have it work. To ensure correct installations, it is important for a job supervisor to explain clearly to workers the purpose of the tasks they perform.

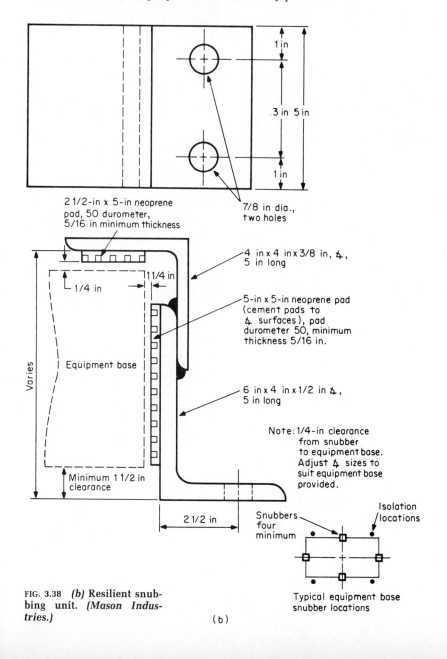

1 in

3 in 5 in

1 in

2 1/2-in x 5-in neoprene pad, 50 durometer, 5/16 in minimum thickness

7/8 in dia., two holes

4 in x 4 in x 3/8 in, ⌊, 5 in long

11/4 in

1/4 in

5-in x 5-in neoprene pad (cement pads to ⌊ surfaces), pad durometer 50, minimum thickness 5/16 in.

Equipment base

6 in x 4 in x 1/2 in ⌊, 5 in long

Varies

Note: 1/4-in clearance from snubber to equipment base. Adjust ⌊ sizes to suit equipment base provided.

Minimum 1 1/2 in clearance

2 1/2 in

Snubbers four minimum

Isolation locations

FIG. 3.38 *(b)* **Resilient snubbing unit.** *(Mason Industries.)*

Typical equipment base snubber locations

(b)

The details shown in Figure 3.38a, if constructed and installed as shown, will prevent the malfunction shown in Figure 3.37. When rigid hold-down bolts like those shown in Figure 3.38a are used, they should be treated acoustically as shown in Figure 3.33. If a resilient separation is not provided at hold-down connections, the high-frequency vibrations that travel down the steel spring coil can go from the coil to the isolator base plate into the hold-down bolt and then into the building structure. This may cause problems if nearby spaces are acoustically sensitive, if vibration-sensitive equipment is used, or if acoustically sensitive tests are taking place.

Figure 3.39 illustrates quite clearly the application of open, stable, steel spring isolators cantilevered from a concrete, steel-framed inertia block with resilient snubbing devices for restricting base and equipment motion during start-up. This method is used where high motor starting torque is anticipated and should not be interpreted to be satisfactory for controlling seismic conditions. See Figure 3.38b for recommended details for the resilient snubbing device.

FIG. 3.39   **Springs with resilient snubbing device.**

3.40   Vibration-Isolated Pump

Figure 3.40 shows a close-up view of part of a vibration-isolated pump and its steel support base. The pump is installed in a theater in Manila. A concrete housekeeping pad has been provided, and one of the spring isolators appears dimly in the background, left (see arrow 1).

Suction and discharge piping has been rigidly connected to the pump flanges in accordance with plans and specifications. The acoustical design for this system was to attach the piping directly to equipment without the use of flexible (vibration-isolating) fittings and then to resiliently isolate the pump and its connected piping from the building. The design was followed except that the contractor installed a steel pipe support stanchion under the pump suction and discharge elbows, extending them down from the elbow to the housekeeping pad. This forms a direct rigid path by which the pump vibrational energy bypasses the spring isolation system. One such steel stanchion support is shown in the foreground of Figure 3.40 (see arrow 2).

This problem occurs frequently; other examples of this misapplication appear in this book. What is typically recommended for such installations is that the pipe elbow stanchion supports come to the equipment base and not to the floor or housekeeping pad below.

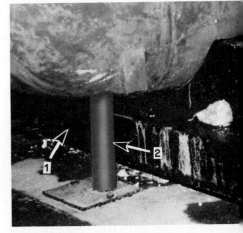

FIG. 3.40   **Vibration-isolated pump.**

The installation shown in Figure 3.41 shows a simple, but very effective, open, stable, steel spring isolator installed in a project in Australia. Here, the spring is nestled in a top and bottom neoprene cup. This isolator unit has been neatly positioned under the equipment frame support base, thus allowing good clearance under the entire isolated package. However, it is apparent that the air-handling unit was placed on temporary steel angle legs tack-welded to the unit base. The equipment, which rested on the temporary legs until the spring isolators arrived on the job, was installed as shown. The completed project included the temporary rigid legs.

The lock screw over the isolator should be bolted tight. Better spring stability and alignment is maintained when this bolt is tightened.

Fortunately, the acoustical consultant on this project was summoned to Hong Kong to check out the sound system. While touring the project, the consultant pointed out the need for the contractor to return to the job and remove the temporary steel legs, thus finally allowing the spring isolators to perform their intended function. Such oversights occur most frequently when no job supervision by appropriate consultants is authorized.

The spring-within-a-spring arrangement, visible if you examine Figure 3.41 closely, is used to afford a greater load-carrying capability while maintaining the same static deflection.

FIG. 3.41   Isolator for air-handling unit.

Air-handling equipment, such as that installed in a
theater project in Australia and shown in Figure 3.42,
can be effectively vibration-isolated on open, stable, steel
springs applied directly under the unit base frame, which
allows excellent clearance underneath the supported
unit. Of course, before the units are finally prepared for
owner acceptance, the springs should be aligned properly. If required, horizontal thrust restraints across the
flexible connections (snubbing devices such as shown on
Figure 3.80) should be installed.

Close examination by the consulting job supervisor
will determine whether the center adjusting bolt is so
long that it can easily contact the isolator base plate. The
bolt can be seen inside the coil of the spring isolator in
the foreground (see arrow). If this is the case, the bolt
must be shortened so that a 1-in clearance is maintained
between the end of the adjusted bolt and the base plate
below it.

Note the several layers of ribbed neoprene pads under
each isolator base plate. These pads are intended to
reduce any high-frequency vibrations that may travel
down the spring coil. Note, too, that there is no need to
bolt the isolators in place in this particular instance. The
pads provide enough frictional resistance to keep the
equipment in place.

FIG. 3.42 **Floor spring isolator for air-handling unit.**

3.43 Isolator for Rooftop Equipment

Spring isolators on outdoor installations should not be covered by the waterproofing used on the supports and curbs.

Although waterproofing or weatherproofing materials are usually soft and remain rather resilient, they should not come into contact with isolators, as is seen in Figure 3.43. Once covered in such a manner, the spring coils cannot be easily seen. Thus the contractor cannot tell if the coils are properly separated from one another and able to operate as they should.

Rather than being waterproofed, the isolators should be specified to be neoprene- or cadmium-coated for environmental protection. In humid climates, as in Puerto Rico where this isolator is located, such protection can significantly lengthen the normal life expectancy of the equipment.

FIG. 3.43   Isolator for rooftop equipment.

3.44 Floor-Supported Equipment
3.45
3.46

There are times when it is acoustically advantageous to support mechanical equipment that would otherwise be hung from above, on elevated platforms or frames from the floor below. Such is the case in the concert hall in Maryland shown in Figure 3.44, which illustrates a pipe column and channel-iron support frame for a large axial-type fan. The support is about 10 ft above the floor with the vibration isolators located on top of the horizontal members of the support and under the supplemental steel fan base. Figure 3.45 is a close-up of one of the isolators. Four such springs were used per fan.

The system being served is a low-pressure, low-velocity return; thus, the isolators were not bolted to the steel frame. Usually bolting, or some other means of securing the isolator base plate, would be necessary only for high-pressure systems, equipment subject to earthquake-resistant requirements, or systems located outside and subject to wind forces.

A very similar method of equipment support and vibration isolation was used in a concert hall in Toronto. The installation can be partially seen in Figure 3.46. Please note in this view that the isolator has not yet been adjusted to pick up its load, so that no clearance shows between the equipment base and steel support members. Once adjusted, a minimum clearance of about 1 in should be maintained.

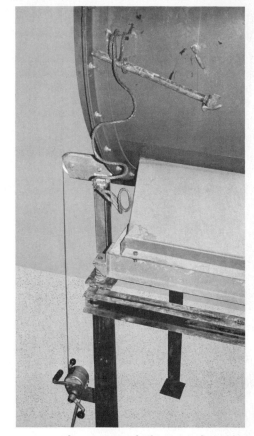

FIG. 3.44   Floor-supported vibration isolation for axial fan.

FIG. 3.45   Open spring isolator for off-the-floor base support.

FIG. 3.46   Double open spring isolator.

3.47   Floor Vibration Isolator

One can see that the spring isolator in Figure 3.47 has not been properly adjusted, because no threads on the bolt are visible below the adjusting nut (see arrow 1).

One very important item that cannot be seen is that the isolator center bolt (see arrow 2) extends so far down inside the spring coils that it actually touches the base plate of the mount. This contact creates a path by which vibrations can travel right into the floor. If this situation is repeated on all the other isolators in this system, it means that the vibrating equipment and its base are really sitting on bolts.

Another small but important detail shown in Figure 3.47 is that, in the field, it is more practical and less troublesome to have the side plates of the angle bracket support cut off at an angle than to use square pieces (see Figures 3.38a, 3.45, and 3.46).

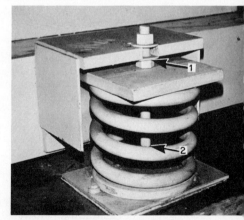

FIG. 3.47   Floor vibration isolator.

3.48 Floor-Mounted Pipe Isolator

The arrangement in Figure 3.48 might be called a vibration isolator supported by pipe stanchions.

In this case, the vertical limit stops have been set improperly. Too much space has been left between the side supports of the isolator and the top plate. The vertical restraining nuts have been run too far down the bolt (see arrow). The hold-down bolts for securing the mount to the pipe stanchion base are far too long and should be cut off to avoid any possible contact with the travel limit bolts. In this arrangement, the top plate of the isolator is really supported only by the center adjustment bolt above the spring. Therefore, the pipe roller is also only supported by this same bolt, and in this arrangement the pipe roller is useless as the spring provides no resistance to the rolling action. If there is expansion, the restraining bolts would be forced against the holes in the isolator housing, and in all probability the top plate would still continue to pivot around the center adjustment bolt.

The installation of pipe rollers on top of spring isolators is seldom successful as the resistance of the spring to horizontal motion is far less than the force required to cause the roller to roll. Expansion of the pipe displaces the upper isolator housing and causes short-circuiting of the bolts through the restraining holes.

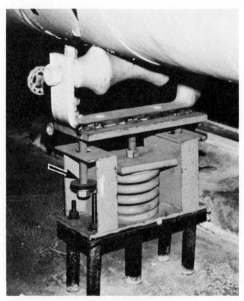

FIG. 3.48   Floor-mounted pipe isolator.

3.49 Floor-Supported Piping Isolation

Figure 3.49 offers additional evidence of what frequently
occurs with a four-point arrangement of the adjusting
bolts. Note the backed-down bolt in the four-spring
isolator (see arrow).

In this installation, the large pipes that occur above
and cannot be seen were recommended not to be hung
from the vibration-sensitive slab above. Thus, floor
supports were framed from structural steel shapes, and
the whole floor-supported system was vibration-isolated
as depicted.

FIG. 3.49   Floor-supported piping isolation.

3.50 Air Spring

Figure 3.50 shows an air spring of the single bag type. A
steel casing is used to provide stabilization for the air
bag. The opening at the casing top allows debris to
collect; the debris may cause damage to the air bag. Units
with open bottoms instead of tops are preferred.

This unit had to be changed in the field because of its
inadequate weight-carrying capability. Regardless of who
is at fault in situations like this, it requires a major effort
to change isolators. Space becomes a very important
factor, and usually the isolated equipment cannot be used
while the replacement is being installed.

FIG. 3.50   Air spring.

3.51 Pneumatic Mounts

3.52 The cylindrical isolators seen in Figures 3.51 and 3.52 are pneumatic mounts or air-spring isolators. Figure 3.51 shows a separately loaded unit; Figure 3.52 shows a group of isolators serving in parallel to provide additional load-carrying capacity. These isolators are inflated, and the inflation is maintained by compressed air connected to an automatic leveling device. The compressed-air tubing may be either copper or polyvinylchloride (PVC). Compression fittings should be used in tubing arrangements. This installation has a control valve for automatic leveling with a built-in time delay to prevent cycling of the air valve. A needle valve in series with an autoleveling valve may be substituted for the time-delay feature. The needle valve may be throttled down to give the same effect as the time delay.

Air springs are normally set in a tripod pattern to ensure stability and avoid the redundancy that may occur with a four-point system. An air dryer for the compressed-air supply should be employed at the compressor to prevent moisture buildup in the air spring. The commercially available rolling lobe air springs shown in Figure 3.51 have a maximum capacity of 5200 lb at 100 psig. If the load at a particular location exceeds this value, use a cluster or series, as shown in Figure 3.52.

Normally, small blocks or brackets dropped below the equipment base are used to provide a maximum ¼-in clearance between the base and the top of the housekeeping pad for safety. In case of air-spring failure, the

FIG. 3.51 **Air spring with snubber.**

equipment can only drop ¼ in, instead of the usual 1 or 2 in required under the rest of the base for housekeeping, short-circuit protection, and venting of the air beneath the support base. Air springs will ultimately wear out, fail, and require replacement because of aging of the neoprene caused by ozone, oil, or other chemicals present in the environment. For convenience of application, an air spring should have the air supply at its base rather than at its top.

In Figures 3.51 and 3.52, the vertical rods are supposedly safety uplift rods, which are to be used in case of overinflation of the air spring and provide vertical seismic protection. Note that the rods in Figure 3.51 (see arrow 1) have a ⅜-in diameter and are attached to a ¼-in-thick base plate; they would fail at a point well below the rated capacity of the air spring, which is 5200 lb. In addition, because of inadequate clearance at the holes in the top of the bracket, these rods create a path along which vibration can travel. Notice that the rod (see arrow) diameters in Figure 3.52 are correct, while those in Figure 3.51 are too small.

The seismic snubber in the foreground in Figure 3.51 is used to prevent horizontal motion only. This bracket is inadequate for the supported loads (approximately 30,000 lb). The bracket attached to the housekeeping pad is slotted (see arrow 2). In effect, this slot can only act as a friction connector, and may come loose during an earthquake. The anchor bolt itself does meet the requirement of Table 26G UBC, 1979, reproduced later as Figure 7.12.

FIG. 3.52   Series air springs.

3.53 Air Springs and Mounts

3.54
3.55 Where very low isolator natural frequencies are required, air springs may be used.

The air spring in Figure 3.53 is a neatly installed rolling lobe type under a cantilevered steel plate bracket, the top plate of which also provides a protective cover over the entire top of the isolator unit. The plate offers protection from leaks that could occur in adjacent piping systems as well as dripping oil or grease applied to system components and equipment during routine maintenance procedures.

This rather unique application consists of air springs (four) installed under a specially constructed steel bed frame. The air springs were selected for very low frequency isolation, i.e., down to 2 Hz. Figure 3.54 shows one of the mounts. The steel bed frame added mass to an otherwise extremely light piece of furniture. The additional weight provided stability so that the air mounts could function without the person in the bed experiencing significant motion.

Air pressure in the air mount is easily controlled and maintained by coupling a bicycle-type hand pump to the air stem connection protruding through the front face of the isolator housing.

The entire system worked extremely well, providing the degree of isolation necessary to allow the householder restful relaxation and good sleeping conditions.

The sources of unwanted vibrations included swimming pool equipment, refrigerators, fans, and miscellaneous electrical equipment.

Figure 3.55 provides illustrations of other types of air springs. The simplest kind of air spring is a piston operating within a cylinder. Such springs were patented and actually used on railroad applications over a century ago. Unfortunately, the sliding seals on the piston not only leaked, but they introduced considerable sliding friction that largely defeated the soft ride they were intended to give.

The nylon-tire-cord-reinforced neoprene rubber spring has no sealing problems and operates with almost no friction. Since the parameters of effective area, volume, and internal air pressure can be varied, a very wide range of characteristics are available for a single air spring.

The important thing to remember is that the rubber does not support the load. The load is supported by a column of air. The spring is a carefully designed package which contains that cushion of air.

(a)

(b)

FIG. 3.53 Air mount: (a) manufacturer's catalog cut and (b) actual installation.

FIG. 3.54 Air-spring mount.

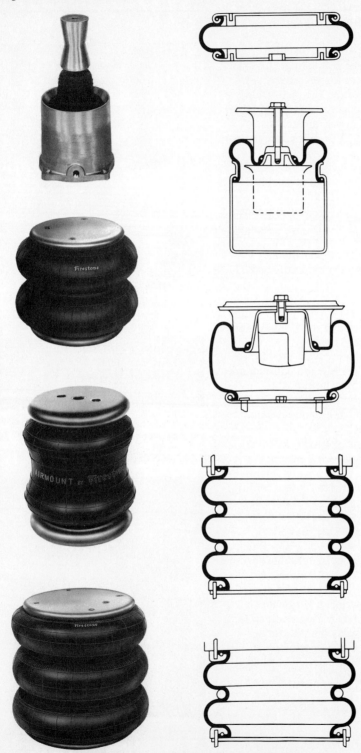

FIG. 3.55   Air springs. *(Firestone Industrial Products Company.)*

DUCTS

SILENCERS AND ENCLOSURES

3.56 Duct Silencers

3.57 Square or rectangular duct silencers, such as those shown in Figures 3.56 and 3.57, are commonly used in ducted air-handling systems to reduce the amount of noise that propagates down the duct from the fan. Noise from fans propagates equally in both supply and return air ducts, regardless of airflow direction.

The preferred method for installing square or rectangular duct silencers in round or oval duct systems is to provide tapered transition sections that gradually change from the shape of the duct to the shape of the silencer, rather than abrupt flat surfaces interfering with the airflow, e.g., those shown in Figures 3.56 and 3.57.

The installations in Figures 3.56 and 3.57 cause considerable reduction in the area of the opening on both the intake and discharge of the silencers. Thus the predicted and certified airflow and acoustical performance ratings of these silencers are reduced.

Generally, according to the Sheet Metal and Air-Con-

FIG. 3.56   Duct silencer at wall penetration.

ditioning Contractors National Association, Inc. (SMACNA), the following practical limits for the design of duct transition sections are recommended.

For high velocity (over 2000 ft/min airspeed in duct), the slope of the sides of the transition section should be a maximum of 1½ in/ft.

For low velocity (up to 2000 ft/min airspeed in duct), a 20° maximum slope for the sides of the transition section for diverging flow; a 30° maximum slope for the sides of the transition section for contracting flow.

Acoustical consultants, as well as duct silencer manufacturers, will generally recommend that duct silencers in noisy mechanical equipment spaces be placed in or as near as possible to the mechanical room wall where the duct leaves the room. Although not commonly understood by many engineers and most installers, the practical reason for this recommendation is as follows.

Mechanical equipment rooms tend to be very noisy because of the simultaneous operation of several pieces of equipment. The noise generated by such equipment also builds or intensifies because of the hard enclosure in which the equipment is usually installed. Typically, these enclosures include concrete floors, concrete block or plaster walls, and concrete slab or metal deck above. Such spaces are highly reverberant; noise "bounces," or reflects, around in the space many times, and very little sound energy is absorbed.

Thus, noise in the room reenters whatever portions of duct lie between the silencers and the point where the duct finally leaves the room. This will happen whether the duct is supplying air from a fan in the mechanical room or is returning air to the fan. The reentry of noise into such exposed duct sections is commonly referred to as *flanking* or bypassing the silencer. One should remember that the primary purpose of the duct silencer is to control the amount of fan-generated noise that is transmitted through the duct to noise-sensitive spaces elsewhere. However, another valuable function that the silencer can perform, if strategically positioned, is to keep out or limit the amount of mechanical room noise.

In cases such as those shown in Figures 3.56 and 3.57, where duct silencers cannot be placed at the wall, floor, or ceiling or a mechanical equipment space and where portions of intervening duct must be exposed to the mechanical room acoustical environment, such exposed duct sections should be acoustically treated. Typically, treatments like those in Figure 3.58 should be used.

FIG. **3.57   Duct silencer in mechanical equipment room.**

3.58 External Duct Treatment _____

3.59
3.60
The usual treatment in this case is to wrap the exterior of the duct section with a minimum of 2-in-thick fiberglass of a density of 2 to 3 lb/ft³ if it is not already thermally insulated. Over the fiberglass, apply two layers of ½- or ⅝-in-thick gypsum board, with all joints staggered and sealed airtight. Acceptable alternatives to the gypsum board are leaded or loaded vinyl or lead with a surface weight of about 1 lb/ft². See Figures 3.58 to 3.60 for further details.

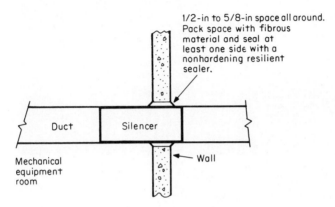

Preferred location

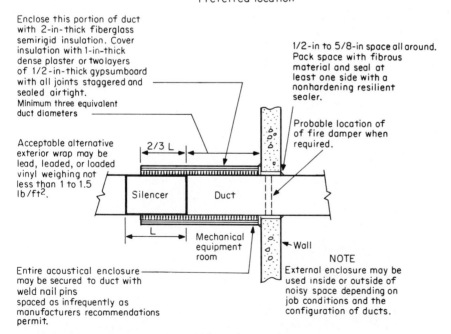

Alternative location

FIG. 3.58 Noise control treatment at duct silencer location.

FIG. 3.59   Duct enclosure with duct silencer.

FIG. 3.60   Duct enclosure in mechanical equipment room.

3.61 Preferred Duct Silencer Location

3.62
3.63
The mechanical consultant for a concert hall in Toronto, Ontario, obtained a waiver of the standard requirement for locating fire dampers in the wall of this mechanical equipment room. This relaxing of a code requirement made it possible to locate the duct silencers seen in Figure 3.61 right in the wall penetrations. This is the preferred location, as shown in Figure 3.58.

The duct silencers for this project are special, in that they have a high TL casing. *High TL* means a high noise-transmission-loss capability for noise entering or leaving the silencer. To achieve a high TL, the silencers are constructed with not only the normal sheet-metal casing but an additional outer metal skin, separated by a fiberglass-filled septum. This high TL treatment goes around all four sides of the silencer for its full length.

FIG. 3.61 **Preferred duct silencer location.**

FIG. 3.62 **Closeup view of elbow silencer.**

This special construction provided the degree of fire protection required to permit installation of out-of-the-airstream fire dampers (see Figure 3.63) on the ends of each silencer, either inside or outside the mechanical room. Being able to position each silencer in its own penetration right in the wall ensures maximum noise control and minimized noise flanking. The annular space between the silencer and the wall is packed and sealed as shown in Figure 3.58.

These are all elbow silencers—that is, each has a full 90° turn in the body of the silencer. A special test was conducted by the manufacturer and witnessed by the acoustical consultant to verify acoustical performance.

A clear elbow configuration can be seen in Figure 3.62 showing the interior of one of the elbow silencers.

Conventional

Out-of-airstream types

Type A

Type A damper is for wall openings or for use with low velocity rectangular ductwork. When used with rectangular duct, the damper is the same nominal size as the duct.

Type B

Type B damper is for use with low velocity rectangular duct where it is desirable to keep the blade stack out of the air stream. When the damper is open, the stack of blades is entirely contained within an enclosure above the duct but the damper sleeve width is the same as the duct width.

Type C round

Type C flat-oval

Type C rectangular

All Type C dampers are for use in high velocity ducts. The dampers have collars of fitting size on both sides of the damper sleeve for attaching the duct. When a Type C damper is open, the blade stack and the blade guides are entirely out of the airstream, thus providing a totally unobstructed flow area with a minimum of pressure drop across the damper.

FIG. 3.63   **Fire dampers. *(United Sheet Metal Division of United McGill Corporation.)***

3.64 Incorrect Installations of Duct Silencers
3.65

Figure 3.64 is shown to demonstrate another way by which duct silencers may be flanked or bypassed by unwanted noise. The view here is looking up a masonry shaft that has been acoustically lined with black-faced fiberglass duct liner. The duct silencer to the left has been placed in the floor slab penetration of the shaft to avoid the flanking of emergency diesel generator noise.

However, the diesel generator exhaust pipe has been installed to pass alongside the duct silencer through a framed opening in the floor slab, with no provisions for blocking the noise path from the generator room to spaces above. As a result, the duct silencer as well as the exhaust pipe muffler (not seen in the picture) will be flanked by the noise of the operating generator.

Figure 3.65, which is a different view of the same installation shown in Figure 3.64, is used here to show another misapplication that further reduces the sound-attenuating capability of the duct silencer. In an effort to get combustion air to the diesel generator without having to provide another acoustically treated penetration in the floor slab above, the contractor installed the combustion air duct through the air passage of one of the silencers. This arrangement reduced the free area of the silencer and covered valuable sound-absorbing interior surfaces, thus reducing the silencer's dynamic insertion loss rating.

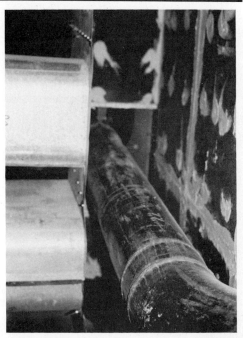

FIG. 3.64 Diesel exhaust pipe opening bypassing duct silencer.

FIG. 3.65 Small vent duct running through duct silencer passage.

3.67 Figure 3.66 is used to show the application of a standard rectangular silencer attached to the intake of an air-handling unit suspended openly in a teaching space.

Figure 3.67 shows a typical conical silencer applied to the discharge side of an air-handling unit. This specific type of duct silencer is fabricated so that the perimeter sound-absorbing media extends beyond the diameter of the attaching ductwork. This explains why the silencer itself is so much larger in diameter than the connecting ductwork. Other types of silencers having diameters equivalent to the ductwork may be used for certain applications where air volume and airspeeds permit.

Besides the need to reduce fan airborne noise on both the intake and discharge of these air-handling units, it was also necessary to control unit-casing-radiated noise. This was accomplished by adding fiberglass insulation sheets (clearly seen in Figures 3.66 and 3.67), over which was applied a final covering of sheet lead, approximately $\frac{1}{16}$ in thick.

The motor and belt drive were also treated acoustically with a sheet-metal, fiberglass-lined enclosure, seen particularly well in Figure 3.67.

Please note the small duct connection from the air-handling unit supply duct, to ventilate the motor enclosure (see arrow). See Figures 3.183 and 3.189 for additional details.

In such applications care must be taken particularly with regard to the air-handling unit intake silencer (Figure 3.66) to check the self-noise characteristics of the silencer to make sure it does not exceed the acoustical design requirement for the space.

In many cases where the silencer self-noise might exceed the noise criteria (NC) levels for the space, by a few decibels, an acoustically lined 90° duct elbow may be added to the silencer intake. The elbow will reduce silencer self-noise transmission by as much as 12 to 18 dB in the frequencies from 250 to 8000 Hz if the duct is at least 40 in wide. Less width will provide less attenuation, and, of course, less attenuation in any case is attainable in the frequencies below 250 Hz. One other important design feature should not be overlooked when using 90° acoustically lined elbows, namely, the legs of the elbow should be sufficiently long so that direct line of sight through the elbow is not possible.

FIG. 3.66 Duct silencer applied to intake of air-handling unit.

FIG. 3.67 Conical duct silencer applied to discharge of air-handling unit.

3.69 A rather popular and practical means of connecting sections of ductwork has come on the market in the form of prefabricated components. These components, generally roll-formed from galvanized steel, may be easily and quickly assembled and, when attached, form a tighter and stiffer joint than usually achieved with older conventional methods.

An installation detail that needs supervision during construction in the field, or at jobsite, is shown in Figure 3.68. A factory-fabricated duct silencer (upper portion of the figure) is being connected to an adjacent section of sheet-metal duct. Due to the interior construction of the silencer, it is not possible to get an airtight seal at the four corners (see arrow) with the system being used. Special closure fittings will need to be fabricated for the corners to ensure airtightness as well as noise tightness.

FIG. 3.68 **Duct with unsealed corner joint.**

There are times when it is acoustically advisable to vibration-isolate certain ducts or portions of ducted air-handling systems from the building structure. Such is the case for this duct installed in a concert hall in Canada, shown in Figure 3.69.

The duct is isolated on double-deflection neoprene mounts which occur between the duct and the angle iron support brackets below (see arrow 1). The brackets, in this case, could only be secured to the wall that is common to the duct shaft and the concert hall. This particular section of duct is relatively close to the supply fan and is therefore handling a rather large volume of air (approximately 15,000 ft³/min). At this point, even though air velocity is only 1200 ft/min, some air turbulence occurs, which excites the duct walls and causes vibration and noise, thus the need for isolation. A higher degree of efficiency may be obtained by using spring isolators.

Because of limited space, the ducts, in several instances, needed additional isolation where the duct face nearest to the concert hall wall actually came into direct contact (see arrow 2). Pieces of neoprene pad were cemented in place between duct and wall at all necessary locations.

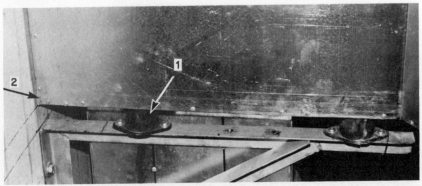

FIG. 3.69 **Vibration isolation for ductwork.**

3.70 Duct Silencer Acoustics Test Setup _____

3.71
3.72
3.73

Figure 3.70 shows a temporary test setup used to verify previously estimated insertion loss data for elbow duct silencers. Tests were conducted both with and without air flowing. The acoustically treated enclosure (see arrow 1) at the left houses the air-moving fan, and the duct terminates in the hard (reverberant) steel plate chamber (see arrow 2) at the right. These tests were conducted by a duct silencer manufacturer on the manufacturer's own premises to prove that the silencers could conform with specified acoustical requirements for a particular project under construction.

The provision for the in-duct noise generator (see arrow 3) (used with no air flowing) may be seen just to the right of the two oil drums beneath the duct. The elbow duct silencer was placed on the two A-frame supports (see arrow 4) at the right side of the figure. The remaining duct sections continue on into the reverberant chamber. Figures 3.71 and 3.72 show the duct termination and reference sound source (fan) within the chamber.

The opening at the right in Figure 3.71 is the acoustically treated opening for running electric and recording cables in and out of the chamber. The opening at the left is the test duct termination. Sound levels were measured with an unlined duct elbow in place with and without air flowing. The measurements were then repeated with the duct silencer inserted in place of the bare elbow.

Microphones were placed in the duct, before and behind the silencer, as well as in the test chamber. The microphones were attached to a graphic level recorder so that all sound levels were recorded, for purposes of comparison as well as for determining the silencer static and dynamic insertion loss (Figure 3.73). Verification of measured sound levels in one-third octave bands was within 1 to 2 dB when compared with extrapolated data from a previous test of ducts and silencers of similar sizes.

FIG. 3.70   Setup for testing duct silencer insertion loss (see Figure 3.73 for schematic layout of test system).

FIG. 3.71   Sound source in calibrated room.

FIG. 3.72   Duct silencer test room showing duct termination.

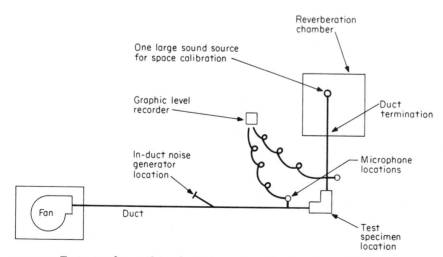

FIG. 3.73   Test setup for verifying duct silencer insertion loss (no scale).

DUCT PENETRATIONS AND CONNECTIONS

3.74 Duct Penetrating Wall

3.75 Penetrations of walls, floors, and ceilings by ducts, pipes, and conduit should be treated acoustically. Figure 3.74 shows that a sheet-metal duct can be correctly centered in a building wall penetration so that acoustical treatment can be installed. See Figure 3.75 for the correct acoustical treatment for duct penetration. At the far right of Figure 3.74 the penetrating duct is shown, capped for future extension. Between the duct and the wall construction, a clear space of about ⅝ in (see arrow) all around has been left for installation of the acoustically resilient separations.

Acoustical treatment at each penetration in a building's structure will produce the best noise and vibration control.

FIG. 3.74 **Duct penetrating wall.**

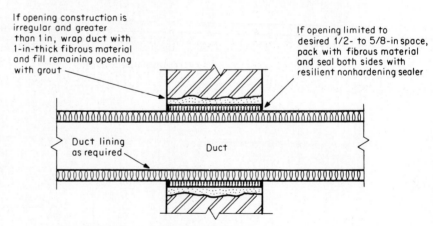

If opening construction is irregular and greater than 1 in, wrap duct with 1-in-thick fibrous material and fill remaining opening with grout

If opening limited to desired 1/2- to 5/8-in space, pack with fibrous material and seal both sides with resilient nonhardening sealer

Duct lining as required

Duct

FIG. 3.75 **Detail of wall penetrations for ducts.**

3.77 Many times in actual constructions on projects, oversized
openings occur for pipes, ducts, and conduit. In Figures
3.76 and 3.77, this situation is shown for ducts that are
penetrating a mechanical equipment room wall on their
way to serve occupied and noise-sensitive spaces in a
building.

Normally, if the space around such a penetrating
member is only ½ to ⅝ in wide, then the acoustical treatment shown on Figure 3.75 is recommended and is
sufficient to control noise transmission and provide resilient separation between the vibrating system and the
building structure.

However, when such annular space is more than the
normal, as described above, and in some cases reaches
proportions of several inches, which is the case in
Figures 3.76 and 3.77, then other methods of closure
need to be applied. Such means are shown and described
in Figure 3.75.

FIG. 3.76   Oversized duct penetration.

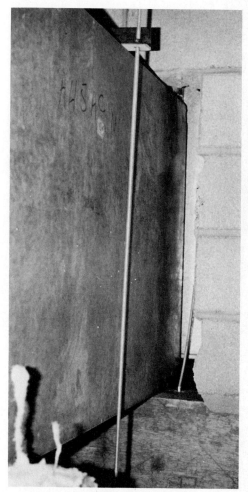

FIG. 3.77   Oversized duct penetration.

3.79 The outdoor duct section in Figure 3.78 was fitted with a
flexible duct connection for vibration isolation. The
connection worked well until the contractor installed a
rigid sheet-metal rain hood (weathercap) across the top
joint. The weathercap destroyed the effectiveness of the
flexible isolation joint.

A recommended treatment for this type of installation
is shown in Figure 3.79. One end of the sheet-metal rain
hood is fastened to the duct. The other end rests on a
foam neoprene strip, which prevents contact between the
duct and the rain hood. The foam neoprene strip is
secured to the duct only, outside the rain hood. The strip
may be glued to the duct but not to the rain hood.

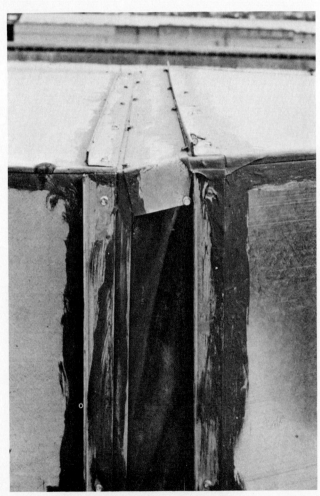

FIG. 3.78 **Duct with rain hood.**

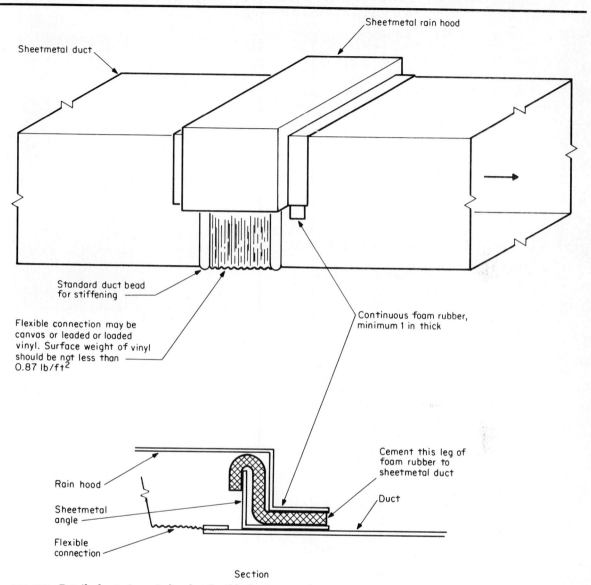

Sheetmetal rain hood

Sheetmetal duct

Standard duct bead for stiffening

Flexible connection may be canvas or leaded or loaded vinyl. Surface weight of vinyl should be not less than 0.87 lb/ft²

Continuous foam rubber, minimum 1 in thick

Cement this leg of foam rubber to sheetmetal duct

Duct

Rain hood

Sheetmetal angle

Flexible connection

Section

FIG. 3.79 Detail of exterior rain hood at flexible duct connection.

3.80 Horizontal Thrust Restraint

Horizontal thrust restraints are effectively employed on the piece of air-handling equipment in Figure 3.80 to resist fan discharge and suction thrusts. However, the improperly installed weather hood (see arrow) over the flexible duct connection contributes to shorting out the resilient thrust restraints. A recommended method for protecting flexible connections installed out-of-doors may be seen in Figure 3.79.

FIG. 3.80 Horizontal thrust restraint.

3.81 Flexible Duct Connection

When properly installed, flexible connections between fans and ducts help to prevent the transmission of vibration. For flexible connections to function as required, proper alignment of components is vital. The improper alignment of fan and duct is shown in Figure 3.81. Vibration energy here can travel from fan to duct because these two components have been allowed to touch one another (see arrow). A 3- or 4-in gap all the way around the connection is generally recommended between the fan and the duct.

Because the treated canvas or glass cloth, which is most commonly used for flexible connections, has a very low surface weight, these connections provide little noise transmission loss. Therefore, they may represent significant leaks of fan noise from the duct.

To contain noise within the fan and connecting duct systems, a material having a surface weight of at least 1 lb/ft² is recommended. Such material could be leaded or loaded vinyl sheet, which is produced in several thicknesses and types to suit a variety of needs. One should be careful in selecting materials of this type to ensure that the required degree of vibration isolation is achieved.

Flexible duct connectors may be field-fabricated or may be purchased as a factory-made item.

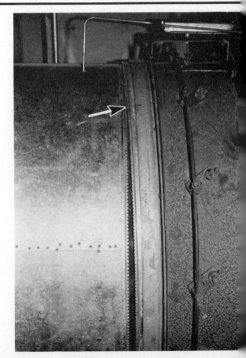

FIG. 3.81 Flexible duct connection.

3.82 Air-Conditioning Duct for Noise-Sensitive Spaces

This large air-conditioning duct (Figure 3.82) was specified to be vibration-isolated from the building structure, with a spring-in-series with neoprene hanger. However, one very important detail about the location of the vibration-isolation hanger is incorrect. The hanger is installed too close to the sides of the duct and one of its stiffener angles. The isolator is in direct contact with both the duct and the angle, and the hanger box is forced to touch the bottom section of the hanger rod, permitting unwanted duct vibration to travel directly through the isolation unit into the building structure.

An alternative is to place small spring isolators under the duct support members. The springs should be seated in neoprene cups having extension bushings similar to those used in factory-fabricated hangers to prevent short-circuiting of the hanger rods against the duct support.

This may require making the bottom duct support members long enough to provide space so that isolation units may be installed as shown in Figure 5.21. The units must be far enough away from the duct sides and stiffeners so that the hanger box may rotate horizontally a full 360° without touching the ductwork. As the mechanical systems operate through on and off cycles, vibrational energy causes the hanger boxes to rotate. Therefore, they need a 360° clearance.

Another installation method is to place the hangers just above the tops of the ducts, between the ducts and the construction above. Even in this location, the hanger box should be placed so that it cannot touch any adjacent systems or elements of the building structure.

A third alternative is to place small spring isolators under the duct support members. This method may require additional height that cannot be provided at some installations. Therefore, this treatment is the least desirable for isolating ducts, and is recommended only when the first two methods cannot be used.

FIG. 3.82 Air-conditioning duct isolation in noise-sensitive spaces.

3.83 Duct Shafts and Steel Angle Supports

Figure 3.83 is a view inside a duct and pipe shaft that runs down through a multilevel hospital building from mechanical equipment on the top floor.

At the floor levels, steel angle and channel irons, which support pipes and ducts, extend across the shaft at intervals. The supports are installed directly on the concrete floor beam and extend somewhat into the shaft partition. In this instance, the partition that separates the shaft from the laboratory on the other side is fabricated from 3½-in metal studs, with two layers of ½-in gypsum board secured directly to both sides of the studs. The laboratory contains precision equipment mounted on benches. This equipment, which includes electron microscopes, is extremely sensitive to vibration. In some cases, the lab benches are freestanding; in other cases, they are secured to the wall. Thus, vibration in the duct and pipe systems, which are secured to the angle and channel-iron supports, can easily be transmitted to the wall, the lab tables, and then to the sensitive equipment.

Pipes, ducts, and conduits, which are supported by these steel cross members, must be vibration-isolated in a manner similar to that shown in Figure 3.93, which appears later.

FIG. 3.83 Duct shaft and steel angle supports.

3.85 In Figures 3.84 and 3.85, the situation is similar to that described for Figure 3.83. Here, however, the separating wall is concrete block, and the vertical duct sections are framed with horizontal steel angles. The angles unintentionally provide shelves on which construction debris can lodge and build up. Though the duct has been separated from the wall during construction, it is now connected to the wall by the debris. This condition is one of many that should be called to the attention of the proper installing contractors at least once or twice during the construction process.

Where vibrating ducts penetrate the shaft wall, these penetrations should be acoustically treated as shown earlier in Figure 3.75. Closing off the space between the duct and the wall by grouting or cementing, as shown in Figure 3.85, is not acoustically acceptable. If space allows, an alternative would be to provide flexible duct connections between the trunk duct and the branch takeoffs.

FIG. 3.84 Excess grout-bridging between duct and wall.

FIG. 3.85 Duct grouted rigidly into building wall.

3.86 Internal Acoustical Treatment of Air Duct
3.87 and Access Doors _____

3.88   To enhance performance and minimize erosion of
fiberglass duct liner, air ducts may be acoustically treated
as shown in Figure 3.86. In this treatment, 1-in-thick
fiberglass duct lining is applied to the inside face of the
duct. The inner duct lining is covered with perforated
metal skin (see arrow).

The back of the field-fabricated duct access door
shown in the foreground of Figure 3.86 is similarly
treated, and the entire inside periphery of the access
door is gasketed with a sponge rubber material, so that
once the latches have been secured, the closure is
airtight and noiseproof. It is possible to purchase factory-
made access doors. Both field- and factory-made access
doors are shown in Figure 3.87.

In Figure 3.88, note the sponge rubber gasket around
the periphery of this access door frame in a sheet-metal
duct (see arrow).

When the access door is inserted into its frame and the
cam locks (usually four) are wound tight, the access door
housing is drawn against the sponge rubber gasket,
creating an airtight and noiseproof seal. This construction
method usually works very well and is an acoustically
acceptable installation.

FIG. 3.86   **Internal acoustic treatment of air duct.**

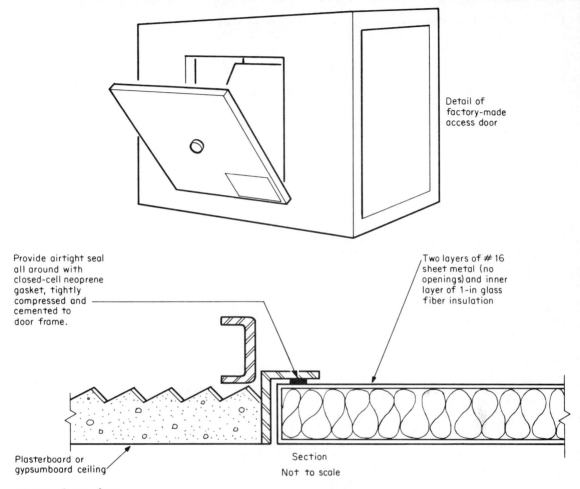

Detail of factory-made access door

Provide airtight seal all around with closed-cell neoprene gasket, tightly compressed and cemented to door frame.

Two layers of # 16 sheet metal (no openings) and inner layer of 1-in glass fiber insulation

Plasterboard or gypsumboard ceiling

Section
Not to scale

FIG. 3.87   Access doors.

FIG. 3.88   Access door for ductwork.

PIPES

SINGLE PIPES

3.89 Single-Pipe Isolation

Figure 3.89 depicts an acceptable method for vibration isolation of piping from the floor or roof of a building. The pipe being isolated here is actually the condenser waterline from an outdoor cooling tower. Thus, the isolators are equipped with travel limit stops to restrict or limit the amount of movement exerted by wind forces acting upon the cooling tower. They also prevent overextension of the springs when the water is drained from the cooling tower.

The steel channel-iron member, which is welded to the condenser pipe, actually supports the pipe. The vibration isolator is located between the channel iron and the steel support system, which either extends down and rests directly on the cooling tower dunnage steel support system or goes to the roof or floor.

The two vertical rods, one on each side of the spring, are the travel limit stops. Note that the nuts have been run down on the threaded part of the travel limit rod to permit the desired degree of movement and to ensure that there is no path along which unwanted vibration can travel.

FIG. 3.89   Single-pipe isolation.

3.90 Condenser Water Main From Cooling Tower

In Figure 3.90, the large pipe seen half embedded in the roof construction of this building in Singapore is one of the condenser water mains from the vibration-isolated cooling tower. It is impractical and ineffective to isolate any piece of vibrating equipment and then counteract the isolation system by joining pipes, drains, city water, conduit, and other such connections directly to the building structure. In such cases, the vibrations transmitted by the equipment travel freely along these direct connections to the building structure, creating noise and vibration problems in acoustically sensitive areas of the building. They may also interfere with vibration-sensitive equipment.

Cooling towers and the condenser water piping connected to them should be vibration-isolated as a total system. Certain cooling tower installations, such as those in remote locations on separate slab at grade, may require no isolation. But even in these installations, portions of the condenser water piping system that enter buildings with noise-sensitive areas will undoubtedly require noise and vibration control treatment.

FIG. 3.90   Condenser water main from cooling tower.

3.91 Pipe Penetrating Building Structure

3.92
3.93
3.94

The space provided between the building structure and the penetrating members of vibrating systems should be packed and sealed acoustically. However, when a serious case of faulty alignment occurs, such as the one shown in this picture, the standard recommended method for installing pipe should be used. See Figure 3.92 for the proper acoustical treatment for pipe penetrating a wall. This treatment is also appropriate for ducts, conduits, and other utilities that penetrate walls.

In general, if the opening in the structure is irregular and greater than 1 in, wrap the pipe or duct with fibrous material 2 in thick and fill the rest of the opening with grout and pieces of brick or concrete block if necessary. If the opening is limited to the desired (½- to ⅝-in) space, pack it with fibrous material and seal both sides with resilient, nonhardening sealer.

Figure 3.91 shows several holes in the wall near the pipe installation. If these holes will not be used, they should be grouted closed. If they are to be used later, plugs of the same material and thickness as the wall should be installed, and the holes should be sealed airtight.

In Figure 3.91, note that the pipe at the bottom right of the picture is positioned correctly in its wall opening and caulked all around with resilient fibrous material. To complete this installation correctly, run a bead of nonhardening resilient sealer around the penetration.

For acoustical treatment where pipes run vertically through the structure, see Figure 3.93.

One of the best ways to assure that pipe sleeves will be centered with the pipe that passes through the sleeve is to install sleeves as shown in Figure 3.94. This is a rather common practice and generally works out quite well so that the space between the sleeve and pipe can be packed all around and sealed. This method allows the pipe to be kept resiliently separated from the sleeve, which gets either poured in place in concrete walls and slabs or grouted in place in masonry unit or other types of construction.

The piping in Figure 3.94 is hung temporarily with solid threaded steel rods. Later, after the piping system has been hydrostatically tested and is filled with water and insulation has been applied, the vibration-isolation hangers will be cut into each rod and adjusted for the final proper weight load. It is at this time that the piping system may be adjusted so that each pipe penetration centers properly in each sleeve.

Some contractors believe a cost saving may be made by installing precompressed hangers for the various locations during construction. Precompressed hangers are self-releasing when water weight is added to the piping system.

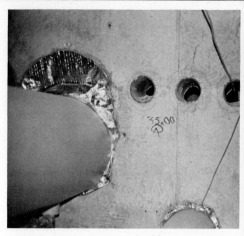

FIG. 3.91   Pipe penetrating building structure.

FIG. 3.94   Pipe with pipe sleeves for wall penetrations.

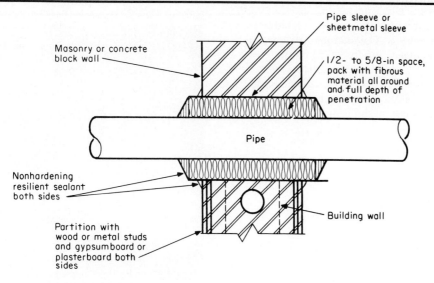

Masonry or concrete block wall

Pipe sleeve or sheetmetal sleeve

1/2- to 5/8-in space, pack with fibrous material all around and full depth of penetration

Pipe

Nonhardening resilient sealant both sides

Building wall

Partition with wood or metal studs and gypsumboard or plasterboard both sides

FIG. 3.92   Detail of pipe penetration.

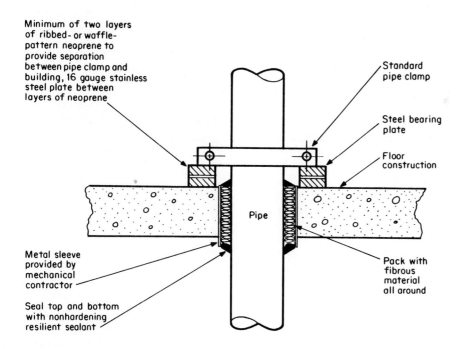

Minimum of two layers of ribbed- or waffle-pattern neoprene to provide separation between pipe clamp and building, 16 gauge stainless steel plate between layers of neoprene

Standard pipe clamp

Steel bearing plate

Floor construction

Pipe

Metal sleeve provided by mechanical contractor

Pack with fibrous material all around

Seal top and bottom with nonhardening resilient sealant

<u>Note</u>:
Neoprene should be 40 to 50 durometer, 40 to 50 psi maximum loading.

FIG. 3.93   Typical floor and ceiling penetration for pipes larger than 1-in diameter.

3.95 Condenser Waterline with Fireproofed Hangers

The waterline in Figure 3.95 is correctly isolated with the proper isolation devices. The example is shown to warn supervising engineers and installing contractors that they should not permit vibration isolators to be fireproofed along with the steel structure of the building. The fireproofing material may significantly reduce the vibration-isolation capacity of the isolator units.

FIG. 3.95 Condenser waterline with "fireproofed" hangers.

3.96 Vibration Isolators for Condenser Water Pipes

Small single-spring vibration isolators with travel limit stops are used in Figure 3.96 to isolate a short run of condenser water piping from a cooling tower. These isolators work well if the travel limit stops are released so that the units function as designed. This picture shows that the nuts under the isolator top plate on both vertical travel limit stop bolts were not released. Considerable vibration was transmitted from the vibrating cooling towers along this piping and into the building, until the nuts were loosened several turns. Loosening the nuts keeps the travel limit stop rods from touching the top plate.

The spring isolators used here are designed to absorb unwanted vibration transmitted through the pipes. The nuts must be released to allow the travel limit stop rods to ride freely through their holes and to release the springs.

When the vibration isolators are locked down, they are designed to be used for blocking under mechanical equipment while it is being installed. Once the equipment is in place, workers often forget that the limit stops must be released so that the isolator can function. A careful job supervisor will catch such errors before the installing contractor leaves the job.

FIG. 3.96 Vibration isolators for condenser water pipes.

This is part of an outdoor installation of condenser water piping to a cooling tower. Because it was out-of-doors, the condenser piping had to be supported above the roof of the building. Some piping sections were several feet above the roof, so the contractor elected to use pipe stanchion supports with the vibration isolators installed in the stanchions, as shown in Figure 3.97. One end of the stanchion is welded to the condenser pipe above, and the other is secured to the roof structure and properly weatherproofed. Both ends of the stanchion are in proper alignment.

The four small pieces of steel bar stock (see arrow), which are welded to the bottom section of the stanchion to keep the base plate of the unit in alignment, are installed correctly. However, they reduce the effectiveness of the ribbed neoprene pad under the base plate of the isolator. Leaving enough space to glue small pieces of neoprene between the bar stock and the isolator base plate would have solved the problem. Instead, the error was repeated over 20 times on this job. The job supervisor should have monitored the contractor's installation step by step to avoid repetition of this error.

FIG. 3.97 **Condenser water piping isolators (floor mounted).**

3.98 Pipe Hanger Embedded in Building Wall

An important step in the building construction process is to coordinate pipe and duct hanger locations with walls, partitions, electrical systems, and other mechanical systems. Figure 3.98 shows a hanger location that cannot be made to work. In this installation, difficulties have been compounded by lack of coordination. The following mistakes make this installation unworkable:

- The pipe itself has been completely cemented into the wall it penetrates, so that vibration from the pipe will pass into the wall.
- The U-shaped clevis hanger used to support the pipe is resting against the wall.
- The vibration-isolation hanger box and the spring itself are touching the wall.
- The two electric conduits that pass below the vibration-isolation unit are pressing against the bottom hanger rod and the top of the clevis hanger.
- The smaller electric conduit near the top of the vibration isolator is touching the hanger box.

Because so many elements of the mechanical and electrical systems have been left untreated, no vibration isolation will be achieved in this installation. Proper job supervision can spot such problems and prevent them, saving the contractor needless repair costs and avoiding embarrassment for the engineer and architect and trouble for the building owner. Timely and well-executed job supervision with good communication between engineers and contractors pays for itself in satisfactory results.

FIG. 3.98 **Pipe hanger embedded in building wall.**

3.99 Heat-Pump-System Water Main with Isolators in Wall

This main is properly hung on the right type of vibration-isolation hanger, but the installation has several obvious problems.

First, the location of the water main was not coordinated with the layouts of the partitions. Thus, the horizontal run of pipe that appears in Figure 3.99 ended up directly above a partition separating a private office from an adjacent section of corridor next to an elevator lobby.

For private offices with acoustical tile ceilings, the acoustical specifications called for partitions extending all the way to the underside of the slab above, to provide the necessary degree of privacy. In an effort to satisfy the requirements of the acoustical specifications, the contractor cut the gypsum board and fitted it carefully around the water main and the vibration-isolation hanger. (The gypsum board has been removed from the left side of the hanger to show the whole installation.)

Because the gypsum board partitions touched the water main, the vibrations in the main caused by the operation of the pump were transmitted into these lightweight partitions. In addition, the main was installed over a suspended ceiling (not shown in the figure), and the gypsum board also touched the ceiling. As a result, the walls and ceiling became radiators of noise and vibration from the pump, causing rattling and vibration in the office.

This problem was not apparent until the office was occupied. Periodic site visits by competent, trained personnel would have caught this error before the construction was completed and the mechanical systems were concealed. Even if the problem were not discovered until the partitions were about to be installed, moving the water main at that point would have been far less expensive than trying to correct the error at the stage shown in Figure 3.99.

FIG. 3.99 Heat-pump-system water main with isolators in wall.

3.102 Figures 3.100 to 3.102 all show vibration isolation for piping using floor mountings instead of the more common method, which is hanging. Floor mounts are used here because an acoustically sensitive space was located directly above this mechanical room. Special acoustical criteria required by the client had to be met.

In Figure 3.100, the large condenser water main is vibration-isolated on open, stable, steel spring mounts. The units have side channels that use travel limit stops to restrict the amount of spring extension occurring when the condenser water main is drained and the weight of water is removed. This installation uses adjustable pipe rolls to support the pipe. The hanger rods are attached to an overhead inverted U-shaped pipe frame with its legs welded to the top plates of individual spring isolators.

The typical inverted U-shaped pipe frame is also used in Figure 3.101, but this installation has a combination steel spring in series with a neoprene element vibration-isolation hanger. Note that the isolation unit is installed upside down. Manufacturers always recommend that the hanger be installed with the neoprene element at the top and the spring at the bottom. The bottom of the hanger box should have a hole large enough to permit the hanger rod to swing through a full 30° arc without touching the sides of the hole. This should be a specification requirement.

The pipe in Figure 3.101 is supported with an adjustable clevis hanger. This means of support may be acceptable if the movement of the pipe from expansion and contraction is limited. In condenser water mains, this movement is usually small.

Figure 3.102 is very similar to the installation in Figure 3.101 except that the vibration-isolation hanger had not been installed at the time of job inspection.

In all three figures, the condenser water pipe was grouted securely into the adjacent wall (penetrated by the pipe main). Thus the effectiveness of the vibration isolators was greatly reduced. The vibrations transmitted to the rigid pipe mains by the chillers and pumps directly connected to them were permitted to travel freely into the concrete block wall and up to the floor of the acoustically sensitive space above. People seated or standing in this space could feel annoying vibrations and could hear excessive reradiated noise from adjacent slab and wall constructions.

FIG. 3.100 **Floor-mounted isolated piping.**

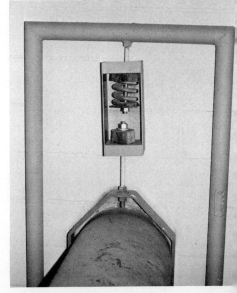

FIG. 3.101 **Inverted U-shaped pipe support.**

The solution to this problem was to chisel out all the grout between the pipe penetrations and the wall, creating a ring of space around the pipe about ½ to ⅝ in wide. The space was packed with fibrous material to the full depth of each penetration and then sealed on both sides with a nonhardening, resilient building sealer (see Figure 3.92).

A manufactured device that may be used as an alternative to the detail shown earlier in Figure 3.92 is shown in Figure 3.103. Such items are quite expensive.

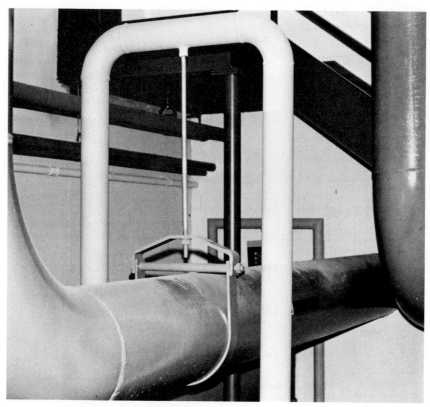

FIG. 3.102   Inverted U-shaped pipe support.

3.103 Pipe Penetration Sealing Device

An inherent feature of this device's flexible rubber body (Figure 3.103) is its ability to absorb shock, sound, or vibration created either by changes in internal pipe pressures or by ground disturbances such as earthquakes or shock waves at railway crossings. Failures due to fatigue are greatly reduced at welds, flanges, and threaded connections.

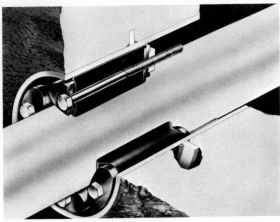

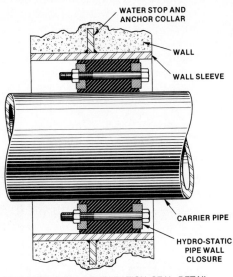

WATER STOP AND ANCHOR COLLAR

WALL

WALL SLEEVE

CARRIER PIPE

HYDRO-STATIC PIPE WALL CLOSURE

PIPE THRU WALL PENETRATION SEAL DETAIL

FIG. 3.103 **Pipe penetration sealing device: *(a) pictorially expressed and (b) detail. (Thunderline Corp.)***

3.104 Vibration-Isolation Hanger

The vibration-isolation hanger in Figure 3.104 has a nut that threads up tight to the bottom of the hanger box. After all piping is aligned and the hangers are installed, the nut is supposed to be loosened to allow space between the hanger box and the equipment to which it is attached. However, installing contractors sometimes forget to loosen the nuts, causing the isolators to remain rigid and therefore ineffective.

Another problem with this hanger is that the hole at the bottom of the box is far too small, permitting only about a ⅛-in clearance between the rod and the hole. The rod protruding through the bottom of the hanger box should be able to swing through a 30° arc before touching the side of the box. The rod must have space to move through a 30° arc to allow for the normal movement in piping and duct systems during system start-up, shutdown, and cycling on and off. If the rod touches the side of the hanger box during these processes, the vibration isolator will not work.

Because of these design problems, the vibration-isolation hanger shown in Figure 3.104 is not recommended and should not be approved. The recommended type of vibration-isolation hanger is shown in Figure 3.5.

FIG. 3.104 **Vibration-isolation hanger.**

3.105 Vibration-Isolation Hanger

The spring-in-series hanger with a neoprene element shown in Figure 3.105 is the proper type and is correctly installed. The isolator is clear of all nearby building construction and other utility systems. The unit may be easily inspected and adjusted; when installed correctly, it will function adequately.

FIG. 3.105   Vibration-isolation hanger.

3.106 Vibration-Isolation Hanger Rods

Figure 3.106 is a graphic demonstration of the on-the-spot ingenuity of the installing contractor. Although it is not standard practice, the contractor shaped hanger rods around the interfering utilities and slipped pieces of black rubber hose over the rods to act as isolating mechanisms (see arrow). This installation worked satisfactorily and was not rejected. However, this is not a recommended practice. Better coordination between the heating and plumbing trades would have prevented this interference.

FIG. 3.106   Vibration-isolation hanger rods.

3.107 Vibration Isolation

The value of specifying vibration-isolation hangers with oversized bottom holes to allow a full 30° swing of the bottom rod is illustrated by the poorly installed hanger in Figure 3.107. The bottom hole here is only slightly larger than the rod itself and thus permits only the smallest misalignment before the rod touches the box and renders the spring ineffective. In actual installations, such devices cannot always be kept in perfect alignment. The slightest shift or movement of the supported systems, such as cycling of a pump or fan, will cause the bottom rod to move out of alignment, and valuable isolating capability will be compromised or lost entirely.

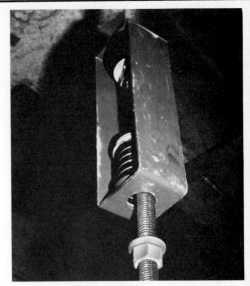

FIG. 3.107 Vibration isolation.

3.108 Vibration-Isolation Hangers

The hangers in Figure 3.108 are identical in type to that shown in Figure 3.107. Here the top hole in each hanger box is small when compared with the rod size, allowing very little movement before rod and box come into direct contact. Both the top neoprene element and the bottom spring can easily be rendered ineffective by a small degree of misalignment. Thus, oversized holes should be specified in vibration isolators for both top and bottom rods.

FIG. 3.108 Vibration-isolation hangers.

3.109 Roller-Type Hanger with Isolators

In Figure 3.109, several items are being hung and vibration-isolated from a trapeze support member. At the far left side of the figure, one end of this support member terminates with a solid hanger rod reaching to the slab above. This arrangement allows the large sheet-metal duct to pass above the trapeze without touching other adjacent systems and without being punctured by hanger rods. If all other installation details have been properly constructed, this method of supporting layered systems works very well. However, note that in this figure the hanger rod touches the side of the hanger box, reducing the device's effectiveness.

FIG. 3.109 Roller-type hanger with isolator.

3.110 Hangers Installed through Duct

In Figure 3.110, no trapeze has been used. Holes have been cut in the duct, allowing valuable conditioned air to leak out. If the isolators are even slightly excited by system vibration, such vibrational energy will be transmitted to the duct, which is not isolated. The vibrations can then travel directly to the building structure. A trapeze hanger arrangement here would have solved the problem without puncturing the duct.

Some incorrect alignment of hanger isolators will occur on almost every project, and there is some leeway allowed for such error. Today's hangers are built to accommodate misalignment up to 15° all around a vertical centerline. These hangers are commonly specified as having a 30° misalignment capability.

FIG. 3.110   Hangers installed through duct.

3.111 Misaligned Hanger Box

Figure 3.111 illustrates what may happen when improper alignment reaches the 30° limit. In this case, the spring cup has been loosened from its proper location, and the neoprene sleeve has been forced out of the hanger box hole. Thus the lower rod has been pressed against the side of the hole in the hanger box, so that vibration can travel through the hanger box.

FIG. 3.111   Misaligned hanger box.

The spring element in Figure 3.112 is totally collapsed
from excessive weight. The installer should immediately
recognize that something is wrong and should remedy
the situation. However, experience continues to show
that without someone to supervise the installation and
order the correction, thousands of improperly installed
units such as this one will be allowed to remain in job
after job, contributing to noise and vibration problems in
buildings.

When vibration-isolation hanger units that are applied
to control the transmission of vibrational energy from a
vibrating system come into contact with static nearby
nonvibrating systems, problems can develop.

FIG. 3.112 **Overloaded spring hanger.**

3.113 Improperly Placed Hangers

In Figure 3.113, at least four isolator hangers have been placed so that after the final installation of several systems, the hangers have come to rest against the side of a sheet-metal duct. In this arrangement, the bottom rods are pushed out of alignment, incapacitating the spring element. The energy from the vibrating system will now pass up the bottom rod into the hanger box, to the lightweight duct, and then into the building structure via the solid duct hangers.

Note that the bottom nuts under the hanger boxes of the two hangers to the left, and of the one in the foreground, have never been loosened to permit the spring element to accept and isolate its load.

In such cases, the following arguments are usually made. The mechanical contractor will say, "We installed the hangers and then the sheet-metal contractor ran the duct." The sheet-metal contractor in turn will profess, "I installed the duct and then the mechanical contractor came and stuck in the hangers." Of course, either argument may be true. Nonetheless, it is clear that someone failed to coordinate the installation, and neither worker cared enough to take steps to correct the situation so that the vibration isolators would function well.

If such jobs are not supervised (usually by a third party), improper installation will occur and remain until the problem of vibration transmission becomes so severe that an acoustical consultant is called in to solve the problem.

FIG. 3.113 Improperly placed hangers.

3.114 Overloaded Hanger

Figure 3.114 shows what happens when isolators are overloaded. Not only does the neoprene bulge, losing its isolating capability in the process, but also the spring coils collapse. In so doing, the spring coils are compressed until they lose their resiliency, thus providing no isolation whatsoever.

FIG. 3.114   Overloaded hanger.

3.115 Vibration-Isolation Hangers

For the record and to offer proof that it can be done correctly, Figure 3.115 shows a couple of vibration-isolation hangers properly installed, perfectly aligned, and apparently selected for just the right weight loading.

FIG. 3.115   Vibration-isolation hangers.

3.116 Floor-Mounted Pipe Elbow Isolator

It is not always necessary to provide the degree of vibration isolation shown in Figure 3.116. However, in this case, because of the immense size of the cooling towers and the huge 20-in-diameter condenser water pipes located just one floor above executive offices in a large high-rise office building, rather special vibration isolation was required.

By design, all condenser waterlines were directly connected to the vibrating cooling towers without using flexible pipe fittings. Therefore, a considerable amount of tower vibration was transmitted along the pipe walls of the condenser water mains between the towers and the refrigeration machines. Thus, the piping was isolated in the fashion shown in Figure 3.116. The isolation includes a small concrete inertia block, point-isolated at its four corners with open, stable, steel springs. The inertia block was used to lower the center of gravity of the assembled pipe section and provide added stability.

One of the large cooling tower isolators is partly visible in the background behind the condenser water pipe. The small spring hanger also seen in the background was installed to vibration-isolate a smaller drainpipe.

FIG. 3.116   Floor-mounted pipe elbow isolator.

3.117 Incorrect Pipe Penetration

Packing and sealing for noise and vibration control treatment around ducts, pipes, and conduits does not necessarily mean cementing or plastering such utilities into the building construction.

If the utility system is connected to vibrating equipment, and the system is connected to walls, floors, ceilings, and columns, the vibrational energy can then be transmitted from the system to the building. In most cases, the vibrations then easily excite building components, especially lightweight walls, partitions, and ceilings. The noise then generated can be annoying and frequently detrimental to the intended use of a space.

The pipe cemented into the wall in Figure 3.117 is coupled to a sewage ejector pump. When the pump operates, a significant amount of vibrational energy is put into this pipeline. The pipeline itself is isolated on spring hangers and is thus separated from the building except where the pipe and part of the mechanical joint fitting are cemented into the wall. An approved method for acoustically treating pipe penetrations is detailed in Figures 3.92 and 3.93.

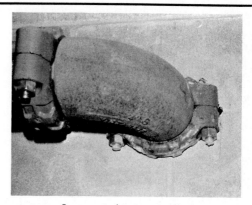

FIG. 3.117   Incorrect pipe penetration.

3.118 Corrected Pipe Penetration

Figure 3.118 shows how the contractor cleaned out around the pipe penetration so that the pipe and the fitting clear the building construction with some room left in which to apply acoustical packing and sealing.

FIG. 3.118   Corrected pipe penetration.

3.119 Flexible Vibration-Isolation Fitting at Pump

The use of steel control rods installed across flexible pipe connectors can often present problems such as is shown in Figure 3.119. Control rods are sometimes required to limit elongation of flexible connectors.

To be totally effective, the control rods must be installed in such a way as to provide no paths by which vibration can transmit across the resilient element of flexible pipe connector. Thus the rods (generally two) must be resiliently isolated where the rods pass through the flange extension hole (see arrow). Often a neoprene sleeve is used at these locations. In addition, resilient washers, usually neoprene, must be used between the flange extension pieces and the holding bolt head on one end plus the threaded nut on the other end.

Misalignment between the pipe sections can easily cause the isolated components to come in direct contact with one another, even when resilient separators are used.

Steel cables are sometimes used in place of control rods, but even these when stretched taut, permit some vibration to pass by the flexible connector.

It is recommended, when using flexible pipe connectors, that they be selected to function under the anticipated pressure and temperature limits without control rods or cables.

Special nitrile liners can be specified for these connectors for fuel oil applications.

FIG. 3.119   Flexible vibration-isolation fitting at pump.

3.120 Recommended Configurations of Steam Piping
3.121 for Minimizing Noise Generation ─────────────────

3.122
3.123
Pipeline noise can be a problem, especially if noisy pipes run in or near work areas or other acoustically critical spaces where quiet conditions are required.

Noise and vibration in piping systems can originate from pumps, compressors, chillers, steam and water pressure-reducing and control valves and other equipment as well as flow-generated noise.

Such noise and vibration can be carried along the pipe or conduit walls as well as by the fluid medium being transported through the pipe.

A common noise source occurring in many buildings drawing heat from a high-pressure steam distribution system can be the steam pressure-reducing valve(s) and the other valves and fittings that are used to fabricate the complete reducing assembly. Such assemblies are commonly called *pressure-reducing stations*. The basic recommended configurations and pipe sizes for typical single- and two-stage reducing stations are shown in Figures 3.120 to 3.123. (Also refer to *Control System Guide* no. 107, Leslie Company, Parsippany, New Jersey, and *Sound Symposium*, also by Leslie Company.) The piping configurations given in Figure 3.120 will generally give good results in minimizing and controlling pipeline noise.

Mathematical expressions and tables for predicting noise in valves and pipes due to aerodynamic and hydrodynamic flows may also be found in the *Masoneilan Noise Control Manual*, Bulletin 023000E. A guide for selecting quiet reducing valves is available from Spence Engineering Co., Bulletin 2500. In addition, *Control Valve Noise*, Bulletin TD850 1a, has been published by Honeywell, Inc., and *Fisher Controls Noise Abatement*, Section 3, contains additional data and selection tables.

Most manufacturers of pressure-reducing and control valves provide valve trim, or in other words, internal valve components designed to minimize noise generated through the valve. The trim has been labeled using terms such as "quiet" and "whisper." Such terms should not be understood to mean that inaudibility has been achieved and thus no more noise problems can occur. Each application should still be considered from the standpoint of how much noise can be expected and how the noise source or sources relate to the desired acoustical environment required in adjacent spaces or spaces through which the pipeline or pipelines pass.

Single Stage Reduction

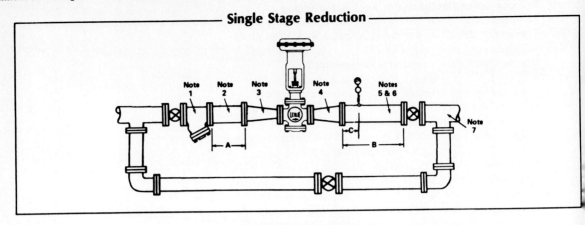

Two Stage Reductions

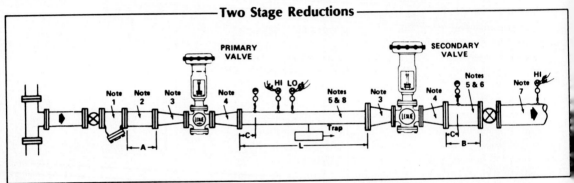

Dimension, ft

Valve Size, in	A	B	C	L (See Note 8)
½–1½	1½	3	2	20
2–4	3	5	3	20
5–8	4	8	3	20

Notes:

1. Size strainer same as inlet piping.
2. Straight run of pipe after strainer of equal size — this dimension may be eliminated if noise is not a factor.
3.* Swage down to valve through tapered fitting of 15 to 20° included angle.

4.* Taper expander (included angle of 15 to 20°) right out of reducing valve outlet to recommended downstream pipe.
5. Straight run of pipe after expander and before outlet stop valve for impulse connection.
6. If downstream pipe is so large that the ratio of pipe diameter to valve size is greater than 2.5 or 3, install a section of straight pipe, twice the valve size, with tapered expander. This section of pipe should be equal to B in the above table. Expanders should have a 15 to 20° included angle. Make control pipe connection, in the first expanded section, at least 2 ft from end of the expander and no closer than 12 to 18 in to the outlet stop valve. Final expansion should be made beyond stop valve with tapered expander.
7. Avoid abrupt change in flow direction.
8. A shorter L dimension is permissible if the diameter of this section of pipe is in accordance with Figures 3.121 to 3.123. However, an L dimension of less than 14 ft is not recommended.

---

* It is possible to make these expansions by using standard fittings and the angle and length can vary by the number of fittings and the ratio used due to standard face-to-face dimensions. This doesn't give quite as smooth an interior and increases the possibility of sharp edges and gasket projections. Also, the angle might be discontinuous if sections are not picked properly.

FIG. 3.120 **Stream pressure reducing stations.** *(Leslie Company.)*

FIG. 3.121 **Minimum Recommended Pipe Sizes for Intermediate Pipe Between Valves**

CHART A   With _14 FEET OF PIPE_ use this chart

Max. Sat. Steam Flow Lbs./Hr.	INTERMEDIATE LINE PRESSURE IN PSIG											
	5	10	20	30	40	50	60	70	80	90	100	125
1,000	3	3	2-1/2	2	2	1-1/2	1-1/2	1-1/2	1-1/2	1-1/4	1-1/4	1-1/4
2,000	4	4	4	3	2 1/2	2-1/2	2-1/2	2	2	2	2	1-1/2
3,000	5	5	4	4	3	3	3	2-1/2	2-1/2	2-1/2	2-1/2	2
4,000	6	6	5	4	4	4	3	3	3	3	2-1/2	2-1/2
5,000	6	6	5	5	4	4	4	4	3	3	3	2-1/2
6,000	8	6	6	5	5	4	4	4	4	4	3	3
8,000	8	8	6	6	5	5	5	4	4	4	4	4
10,000	10	8	8	6	6	5	5	5	5	4	4	4
12,000	10	10	8	8	6	6	6	5	5	5	5	4
14,000	12	10	8	8	8	6	6	6	5	5	5	4
16,000	12	10	10	8	8	6	6	6	6	5	5	5
18,000	12	12	10	8	8	8	6	6	6	6	5	5
20,000	14	12	10	10	8	8	8	6	6	6	6	5
25,000	16	14	12	10	10	8	8	8	8	6	6	6
30,000	16	16	14	12	10	10	8	8	8	8	8	6
35,000	18	16	14	12	10	10	10	8	8	8	8	8
40,000	20	18	14	12	12	10	10	10	8	8	8	8
45,000	20	18	16	14	12	10	10	10	10	8	8	8
50,000	20	20	16	14	12	12	10	10	10	10	10	8
55,000	24	20	18	16	14	12	12	10	10	10	10	8
60,000	24	20	18	16	14	12	12	10	10	10	10	10
65,000	24	24	20	18	16	14	12	12	12	10	10	10
70,000	24	24	20	18	16	14	14	12	12	12	10	10
75,000	24	24	20	18	16	16	14	14	12	12	12	10
80,000	–	24	20	18	18	16	16	14	14	12	12	10
85,000	–	24	20	20	18	16	16	14	14	12	12	10
90,000	–	24	24	20	18	16	16	14	14	14	12	10
95,000	–	24	24	20	18	18	16	16	14	14	12	12
100,000	–	–	24	20	20	18	16	16	16	14	14	12
125,000	–	–	24	24	20	20	18	18	18	16	16	12

Line velocity will be limited to 100-135 ft./sec. (6,000-8,000 ft./min.)

Source: Leslie Company

FIG. 3.122 **Minimum Recommended Pipe Sizes for Intermediate Pipe Between Valves**

CHART B   With _16 FEET OF PIPE_ use this chart

Max. Sat. Steam Flow Lbs./Hr.	INTERMEDIATE LINE PRESSURE IN PSIG											
	5	10	20	30	40	50	60	70	80	90	100	125
1,000	3	2-1/2	2	2	1-1/2	1-1/2	1-1/4	1-1/4	1-1/4	1-1/4	1-1/4	1
2,000	4	4	3	2-1/2	2-1/2	2	2	2	2	1-1/2	1-1/2	
3,000	5	4	4	3	3	2-1/2	2-1/2	2-1/2	2-1/2	2	2	1-1/2
4,000	5	5	4	4	4	3	3	3	2-1/2	2-1/2	2-1/2	2
5,000	6	5	5	4	4	4	3	3	3	2-1/2	2-1/2	2-1/2
6,000	6	6	5	5	4	4	4	4	3	3	3	2-1/2
8,000	8	8	6	5	5	4	4	4	4	4	4	3
10,000	8	8	6	6	5	5	5	4	4	4	4	4
12,000	10	8	8	6	6	5	5	5	4	4	4	4
14,000	10	10	8	8	6	6	5	5	5	5	4	4
16,000	10	10	8	8	6	6	6	5	5	5	5	4
18,000	10	10	8	8	8	6	6	6	5	5	5	5
20,000	12	10	10	8	8	8	6	6	6	5	5	5
25,000	14	12	10	10	8	8	8	6	6	6	6	5
30,000	16	14	12	10	10	8	8	8	8	6	6	6
35,000	16	14	12	10	10	10	8	8	8	8	8	6
40,000	18	16	14	12	10	10	10	8	8	8	8	6
45,000	18	16	14	12	10	10	10	10	8	8	8	8
50,000	20	18	16	14	12	10	10	10	10	8	8	8
55,000	20	18	16	14	12	12	10	10	10	10	8	8
60,000	20	20	18	14	14	12	12	10	10	10	10	8
65,000	24	20	18	16	14	12	12	10	10	10	10	8
70,000	24	20	18	16	14	12	12	12	10	10	10	8
75,000	24	24	20	16	16	14	12	12	10	10	10	10
80,000	24	24	20	16	16	14	14	12	12	10	10	10
85,000	24	24	20	18	16	14	14	12	12	12	10	10
90,000	–	24	20	18	16	14	14	14	12	12	12	10
95,000	–	24	20	18	16	16	14	14	12	12	12	10
100,000	–	24	20	20	18	16	14	14	14	12	12	10
125,000	–	–	24	20	20	18	16	16	14	14	14	12

Line viscosity will be limited to 135-170 ft./sec. (8,000-10,000 ft./min.)

Source: Leslie Company

FIG. 3.123 **Minimum Recommended Pipe Sizes for Intermediate Pipe Between Valves**

CHART C **With _18 FEET OF PIPE_ use this chart**

Max. Sat. Steam Flow Lbs./Hr.	INTERMEDIATE LINE PRESSURE IN PSIG											
	5	10	20	30	40	50	60	70	80	90	100	125
1,000	2-1/2	2	2	1-1/2	1-1/2	1-1/4	1-1/4	1-1/4	1-1/4	1	1	1
2,000	3	3	2-1/2	2-1/2	2	2	2	1-1/2	1-1/2	1-1/2	1-1/2	1-1/4
3,000	4	4	3	3	2-1/2	2-1/2	2-1/2	2	2	2	2	1-1/2
4,000	5	4	3	3	3	3	2-1/2	2-1/2	2-1/2	2	2	2
5,000	6	5	4	3	3	3	3	3	2-1/2	2-1/2	2-1/2	2
6,000	6	5	5	4	3	3	3	3	3	2-1/2	2-1/2	2-1/2
8,000	8	6	5	5	4	4	3	3	3	3	3	2-1/2
10,000	8	8	6	5	5	4	4	4	3	3	3	3
12,000	8	8	6	6	5	5	4	4	4	3	3	3
14,000	10	8	8	6	6	5	5	5	4	4	4	3
16,000	10	8	8	6	6	5	5	5	5	4	4	4
18,000	10	10	8	8	6	6	5	5	5	5	4	4
20,000	10	10	8	8	6	6	6	5	5	5	5	4
25,000	12	10	10	8	8	6	6	6	6	5	5	5
30,000	14	12	10	10	8	8	8	6	6	6	6	6
35,000	14	12	10	10	8	8	8	8	8	8	6	6
40,000	16	14	12	10	10	8	8	8	8	8	8	6
45,000	16	16	12	10	10	10	8	8	8	8	8	6
50,000	18	16	14	12	10	10	10	8	8	8	8	8
55,000	18	18	14	12	10	10	10	10	8	8	8	8
60,000	20	18	16	12	12	10	10	10	8	8	8	8
65,000	20	18	16	14	12	10	10	10	10	8	8	8
70,000	20	20	16	14	12	12	10	10	10	10	8	8
75,000	20	20	16	14	14	12	12	10	10	10	10	8
80,000	24	20	18	16	14	12	12	10	10	10	10	8
85,000	24	20	18	16	14	12	12	12	10	10	10	8
90,000	24	20	18	16	14	14	12	12	12	10	10	8
95,000	24	24	20	16	16	14	12	12	12	10	10	10
100,000	24	24	20	18	16	14	14	12	12	10	10	10
125,000	-	24	20	18	16	16	16	14	14	12	12	10

Line viscosity will be limited to 170-200 ft./sec. (10,000-12,000 ft./min.)

Source: Leslie Company

3.124 Controlling In-Pipe Noise

3.125 The control of pipeline noise may be achieved in various ways depending on the noise path or paths.

Noise transmitted along inside the pipe may be reduced by installing pipe silencers similar to those produced by most of the reducing valve manufacturers. See Figure 3.124a for a typical detail of a pipe noise suppressor and muffling orifice and Figure 3.124b for a detail of an in-line silencer.

In new installations, the noise can be controlled by selecting reducing or control valves with noise-suppressing trim having specific acoustical ratings and by designing and sizing adjacent valves and fittings for the complete assembly or "station" for best flow conditions.

THE SPENCE MUFFLING ORIFICE

View on inlet side   View on outlet side

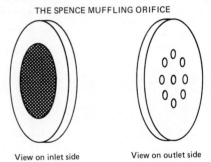

FIG. 3.124a **Pipe noise suppressor and muffling orifice. (*Spence Regulator Co.*)**

Considerable noise can be generated by poor pipe configurations. See Figure 3.125.

Some applications, depending on pressure and temperature requirements, may allow the use of flexible pipe connectors. Flexible pipe connectors serve the function of providing a structural break or separation in the pipeline, thus minimizing the transmission of vibration along the pipe walls. These fittings may also be useful in damping out some pulsations if the medium transported by the pipe is fluid and if the fitting is specifically designed for this additional feature.

Spence noise suppressors provide an economical means of reducing pipeline-transmitted noise by 10 to 20 dBA. These noise suppressors are of the dissipative-

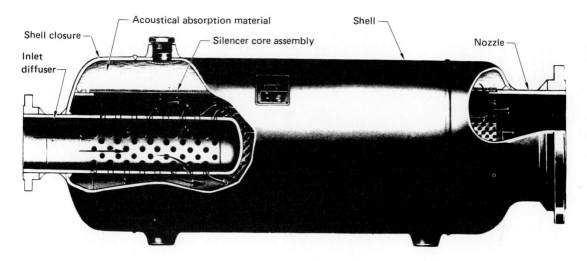

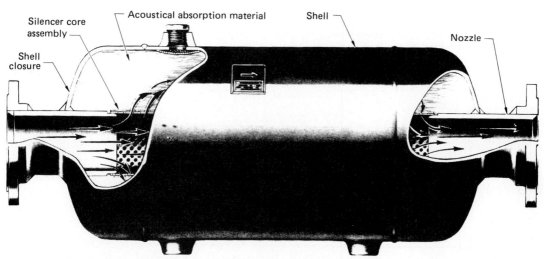

FIG. 3.124b **In-line horizontal silencers.** *(Fisher Controls.)*

reactive type. They are effective over a broad (2,000 to 12,000-Hz) frequency band. Spence noise suppressors are designed to be installed immediately downstream of a pressure-reducing valve. The high-velocity turbulent flow leaving the regulator expands inside the noise suppressor. A large portion of the noise generated by the reducing valve is absorbed and thus prevented from propagating into downstream spaces. The noise suppressor's straight-through construction provides minimal pressure drop and does not affect reducing valve sizing.

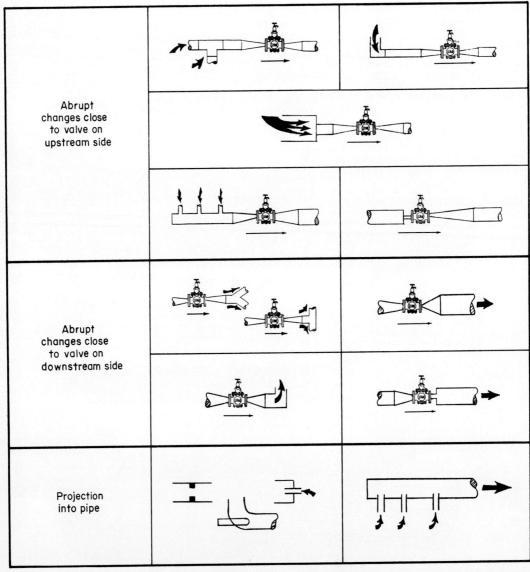

FIG. 3.125   Possible causes for noise in a reducing station. *(Leslie Company.)*

Spence muffling orifices provide an economical means of reducing pressure regulator noise by 6 to 10 dBA. Muffling orifices are individually engineered for each application to provide maximum noise reduction without reducing regulator capacity. Available in 2- through 16-in pipe sizes, the muffling orifice is bolted between standard ANSI pipe flanges. It is installed either in the expanded pipe downstream of the regulator or between the mating flanges of the regulator and the noise suppressor.

Silencers are devices that are installed in the flow path to dissipate the sound energy by absorbing it in an acoustical pack. They are designed to take less than 1-psi pressure drop. In-line silencers are often the most economical approach to noise control where high mass flow rates and low pressure drops exist. These units are normally used immediately downstream of control valves but, in some cases, may also be required upstream of the valves.

When flexible pipe connectors cannot be used or are not preferred, then the piping itself may need to be vibration-isolated from the building with resilient isolators.

3.126   **Controlling Pipe-Radiated Noise**

3.127   Controlling noise from pipes may also be achieved by wrapping the pipeline in a manner similar to that shown in Figures 3.126 and 3.127. The pipe being acoustically treated is a large (18-in-diameter) steam main passing through the basement area of a building. Certain areas of the basement are now being utilized as laboratories and offices, and it was required by the owner of the building that noise levels be reduced.

The external wrapping in this case consisted of (1) a 2-in thickness of fiberglass insulation approximately 3 lb/ft^3 density and (2) a final wrap of lead-loaded vinyl sheeting weighing 1.5 lb/ft^2. All joints, both transverse and longitudinal, were lapped and sealed with a special cement compatible to the sheeting. Patterns were made, in certain locations where careful and concise fitting was required, so that sheeting pieces would fit correctly and overlap the required amount to ensure a complete and uninterrupted cover of all pipe and fittings.

Entire reducing valve stations may be enclosed in an acoustically designed compartment or housing. In such cases, the enclosure will need to be mechanically ventilated to control excessive heat buildup. In addition, adequately sized access doors or panels will need to be provided for proper servicing and maintenance of all enclosed valves, traps, gauges, thermometers, etc.

Some suggestions for specifying piping system flow rates and allowable acceleration levels are given in *Considerations in the Specification of Pipeline Noise*, by Leslie Co.

FIG. 3.126   **Acoustically wrapped steam main.**

FIG. 3.127   **Fiberglass and leaded vinyl cover over steam main.**

3.128 **Controlling Pipeline Noise with**
3.129 **Special Devices** _____
3.130 There are also manufactured devices for controlling
3.131 water hammer* in piped systems, cavitation† in pumped
circuits, and "quiet" or "silent" check valves‡ for
close-off when pumps shut down. See Figures 3.128 to
3.131 for illustrations and principle of operation.

Principle of operation. Flow takes place in the direction
of the arrows as indicated in Figure 3.128. With the flow
in the direction of the arrows, the diaphragm deflects
inward (A) allowing easy passage of the fluid with very
little pressure loss. When a back-pressure or no-flow
condition occurs, the diaphragm immediately resumes its
relaxed position (B) covering the holes in the cone
creating tight shutoff. The check valve can be mounted
in any orientation.

Type CD consists of only two major components: a
metal cone and a flexible diaphragm. It can be mounted
quickly and easily between standard ANSI 125/150/
300 lb flanges.

How Pulsatrol works. Boyle's law $P_1V_1 = P_2V_2$ is the
principle involved in pneumatic-hydraulic pulsation
dampening. The single-diaphragm Pulsatrol has air or gas
sealed under pressure in the upper section of the
chamber. Liquid is isolated in the bottom portion by the
diaphragm. The double-diaphragm model has air or gas
sealed in two opposing chambers. Liquid is isolated
between the two diaphragms.

Pressure changes within the product line due to shock
or surges are instantaneously absorbed by the compress-
ible air or gas volume. Flow changes of a noncompress-
ible fluid in a pipe, acceleration, deceleration, of sudden
stoppage due to rapid closure of a valve, will cause a
sharp pressure change. A Pulsatrol pulsation dampener
absorbs such shocks.

Unlike common pressure dome or surge chamber
devices, the Pulsatrol is fitted with a molded diaphragm
to *isolate* the charged gas chamber from fluid in the
pipeline. This prevents loss of air cushion by absorption
into liquid system.

To operate, it is only necessary to charge the air
chamber to a pressure approximately 50 percent of anticipated mean line pressure. The diaphragm will assume

_____

* Wade shokstops 104 from Tyler Corp. and Pulsatrol Catalog
730 Pulsafeeder Prod./Interpace.
† Fisher Controls Bulletin E-15.
‡ CPV Catalog 690-E, Metroflex Bulletin 900, Leslie Cone
check 5289, and Techno Corp. Bulletin 167.

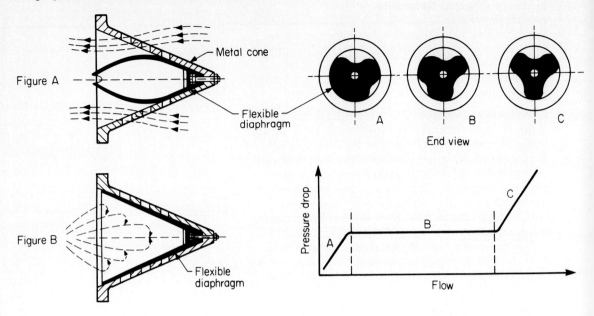

Metal cone

Figure A

Flexible
diaphragm

End view

Figure B

Flexible
diaphragm

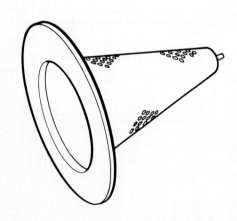

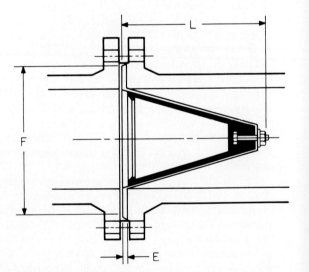

Pipe size, in	E	F	L
1 1/2	5/64	3 7/32	2 1/4
2	5/64	4 1/64	3
2 1/2	5/64	4 41/64	4 1/8
3	5/32	5 3/16	5 3/16
4	5/32	6 7/32	6 3/4
6	5/64	8 11/32	10 5/8
8	5/64	10 9/16	14 3/8
10	5/64	12 19/32	18

FIG. 3.128   Diaphragm-type check valve.

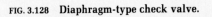

an intermediate position, completely balanced and free to flex as pressure surges occur.

What Pulsatrol does. Pulsatrol smooths out pulsating flow caused by positive displacement piston, cam, or roller-type pumps.

- Eliminates hydraulic hammer in pipelines.
- Reduces peak velocity of the sinusoidal flow characteristics of small reciprocating pumps.
- Establishes more favorable net positive suction head (NPSH) conditions on inlet side of metering pumps.
- Allows use of smaller piping sizes.
- Protects piping and joints from peak pressures and leakage.
- Changes pulsating flow to near linear in combination with back-pressure valve or orifice.

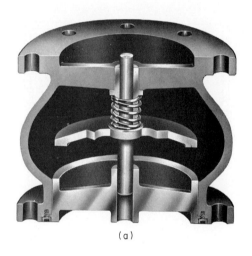

(a)

FIG. 3.129 Quiet-type check valves.
(a) Globe style. *(The Metraflex Co.)*
(b) Metal lined. *(Techno Corp.)*

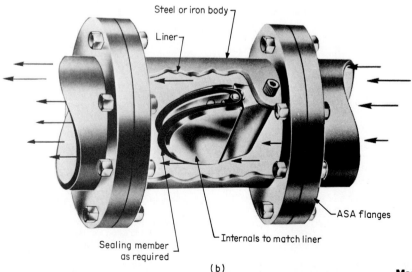

Steel or iron body

Liner

ASA flanges

Internals to match liner

Sealing member
as required

(b)

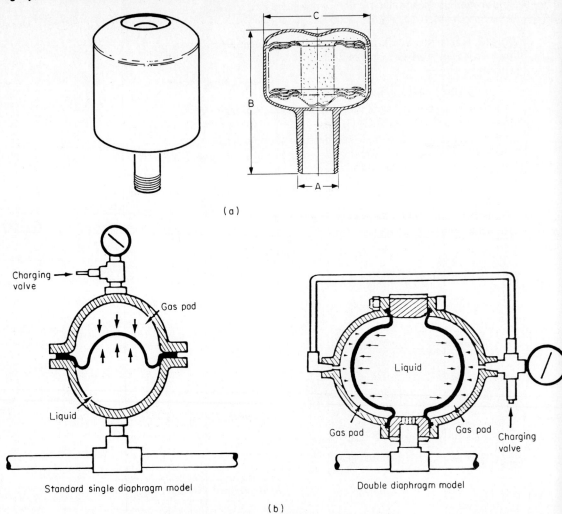

(a)

Charging valve

Gas pad

Liquid

Liquid

Standard single diaphragm model

Gas pad   Gas pad

Liquid

Charging valve

Double diaphragm model

(b)

FIG. 3.130   Pulsation dampers. [(a) Wade. (b) Pulsafeeder Prod.]

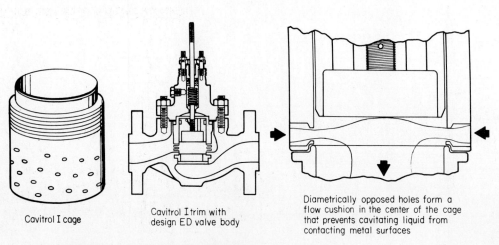

Cavitrol I cage

Cavitrol I trim with
design ED valve body

Diametrically opposed holes form a
flow cushion in the center of the cage
that prevents cavitating liquid from
contacting metal surfaces

FIG. 3.131   Anticavitation device. (Leslie Company.)

MULTIPLE PIPES

3.132 Trapeze Supports for Pipes

Trapeze supports may be used to vibration-isolate several vibrating piping systems at once, as in Figure 3.132. In such installations, care should be taken not to combine vibrating and nonvibrating systems. For example, a pumped system should not be supported with a static sprinkler system or with stationary electrical conduits. Such system mismatches must be avoided.

In Figure 3.132, solid hanger rods were placed at each end of each trapeze channel. The vibration-isolator hangers were cut in later. This method of installation is perfectly acceptable.

Precompressed hanger isolators would be appropriate for this application.

FIG. 3.132   Trapeze supports for pipes.

3.134 The trapeze hanger for pipes shown in Figure 3.133 has been installed too close to the concrete building beam. In this position it was easy for the electrical contractor to run the conduit through the hanger box. The hanger should not be touching either the electric conduit or the building beam. Actually, the box should be located away from the beam, so that it can rotate horizontally the recommended full 360° without touching any part of the structure or adjacent utilities. The electrical conduit should be relocated to a position that does not conflict with the hanger.

A typical trapeze arrangement for multiple conduits where no vibration isolation is required is shown in Figure 3.134. A neoprene sleeve extending through the bottom hole of the hanger box would be most practical.

FIG. 3.133   Trapeze pipe isolator.

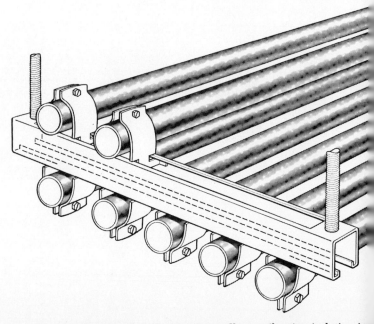

FIG. 3.134   **Multiple-conduit trapeze. Generally no vibration isolation is required unless conduits are directly connected rigidly to vibrating equipment.** *(U.S. Gypsum.)*

3.136   This trapeze support has vibration-isolation units at each end (only one is visible in Figure 3.135). The support was installed properly before construction of the concrete masonry block wall.

In Figure 3.135, the end of the trapeze and its isolator were grouted into the wall, permitting the vibrational energy in the piping systems to travel along the trapeze support member and into the building. The vibration-isolation unit is thus rendered useless.

To correct the situation, the bottom support member of the trapeze was shortened so that it stopped short of the wall. Then the isolation hanger unit was moved into the room, away from the wall, to clear all obstructions and thus keep the entire assembly clear of the building structure. See Figure 3.136 for recommended trapeze isolation system.

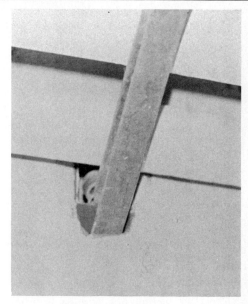

FIG. 3.135   **Trapeze pipe support.**

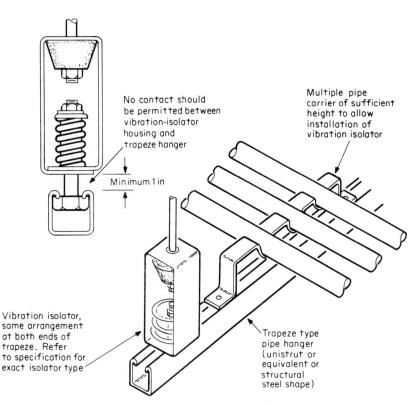

No contact should be permitted between vibration-isolator housing and trapeze hanger

Minimum 1 in

Multiple pipe carrier of sufficient height to allow installation of vibration isolator

Vibration isolator, same arrangement at both ends of trapeze. Refer to specification for exact isolator type

Trapeze type pipe hanger (unistrut or equivalent or structural steel shape)

FIG. 3.136   **Recommended vibration isolation for multiple pipes close to underside of building construction above (no scale).**

113

3.137 Pipe Hanger

Many acceptable methods can be employed to vibration-isolate pipes resting on trapeze support members. Another method is shown in Figure 3.137.

In this case, two steel channels have been welded together, back to back, with steel plate straps positioned to allow just enough space between the channels for the hanger rods to pass.

Cantilevered channel-iron brackets have been fabricated and welded together to form the element that is bolted (or welded) to the building structure and from which the trapeze support members hang. A piece of steel pipe is angle-cut and welded to the side and bottom leg of the wall support bracket to provide the required structural strength (see arrow).

The bracket off the wall was used only because the rod could not travel through to the ceiling above because of the interference of ducts, piping, or other equipment.

The vibration-isolation unit used in this case is a double spring in series with a neoprene element, separated by an intermediate load-distributing platform.

The bottom hole of the hanger box is unnecessarily small in diameter, and a small degree of misalignment could render the hanger useless. Today most manufacturers are offering isolation hangers with oversized holes to help alleviate this common problem.

FIG. 3.137 Pipe hanger.

3.138 Trapeze Isolators for Large Pipes

In Figure 3.138, several large water-carrying pipes are vibration-isolated on a trapeze support, which in turn rests on spring mounts. If the pipes are drained, the removal of the water causes a significant decrease in the weight imposed on the springs. Therefore, travel limit stops are employed to prevent overextension of the springs.

Wall brackets were used due to interferences above that prevented extending rods up to construction.

FIG. 3.138 Trapeze isolators for large pipes.

3.139 Pipe Hangers

There are many ways to hang pipes singly or in groups. Figure 3.139 shows several means of hanging pipes.

In the foreground, a steel channel is cantilevered from the building to provide a trapeze hanger for a group of pipes that must be vibration-isolated from the building. In this case, as well as others in this figure, the vibration isolators are simple double-deflection neoprene elements in steel hanger box. Behind the trapeze hanger, a row of five individual hangers with vibration isolators has been installed, ready to receive its pipe section.

In the left background, two pipes have been installed on their isolators with steel pipe clamps, separated from the pipe with wood blocking for thermal purposes.

FIG. 3.139 **Pipe hanger.**

3.140 Isolation for Cooling Tower and
3.141 Condenser Waterlines

The cooling tower on the roof of this building is well-isolated from the supplemental steel support system, sometimes called *dunnage*. Two of the multispring isolation units are clearly visible in Figure 3.140.

However, the large steel condenser waterlines connected to the cooling tower have no vibration isolation at all (see arrow). Serious vibration problems developed once the cooling tower began to operate. To understand the serious oversight this represents, one might think of these 12-in condenser water pipes as structural steel members. A great deal of vibrational energy will travel along these pipes when they are coupled to a vibrating cooling tower at one end and to a large chiller or refrigeration compressor at the other end, with pumping equipment in between.

Generally, cooling towers, chillers, pumps, and all connecting piping should be vibration-isolated on spring-in-series isolators with neoprene elements. Either floor-mounted or hanger isolators should be used, depending on job conditions. This treatment is preferred because it isolates the entire system both from equipment vibration in the building and from vibration generated by turbulence in the piping system.

However, in some situations and applications, pipe fitting isolators, such as shown in Figure 3.141, may be used to separate piping systems from pieces of vibrating

FIG. 3.140 **Isolation for cooling tower and condenser water lines.**

equipment. The neoprene connectors are fully dimensioned, described, detailed, and sized in the manufacturer's catalog. Several types that may be used are shown in Figure 3.141. If these devices are used, they should be installed in strict accordance with the manufacturer's recommendations. These units have proved very effective in reducing unwanted vibration and noise transmittal at hydraulic frequencies equal to the number of pump impeller blades multiplied by the revolutions per minute.

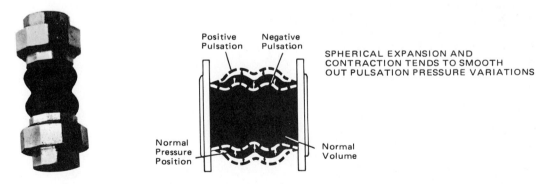

FIG. 3.141 Flexible pipe connectors. *(Mason Industries.)*

3.142 Multiple-Pipe Penetration

A certain degree of privacy, or at least a limit on noise intrusion, is usually desirable between offices in buildings. Such privacy is needed in certain administrative offices in a concert hall in Toronto, Ontario.

Paths that permit flanking of noise around or through partitions or walls between offices include openings like the one seen in Figure 3.142. A group of heating and chilled water mains run along the inside of the exterior wall, penetrating each office wall. If the opening is left acoustically untreated, as in the figure, a cross-talk path will exist. Thus all openings around the penetrating members must be closed and sealed. The most efficient method for this treatment is detailed in Figure 3.92. The correct and complete application of noise control treatment will leave the penetration essentially airtight and noiseproof.

FIG. 3.142 **Multiple-pipe penetration.**

BOILERS

3.143 **Boiler Isolation** _____

3.144
3.145
3.146

In certain instances, when boilers are located either on upper building levels or close to acoustically critical spaces, this equipment and its attachments must be vibration-isolated.

In Figure 3.143, the boiler was isolated on open, stable, steel spring units. These units were provided with travel limit stops to control the amount of spring extension when the boiler was emptied of water.

Gas and electric connections to the boiler were fitted with long, floppy sections of braided flexible tubing and flexible armored electric conduit. Both can be seen in the foreground of Figure 3.143. The city water connection was fitted with a flexible section of pipe, and the boiler drain piping was run to the floor drain without direct connections to the drainage system.

The hot-water heating mains to and from the boiler, as well as the boiler breeching (smoke vent), were hung on spring-in-series with neoprene vibration-isolation hangers.

The boiler shown in Figures 3.144 and 3.145 is point-

FIG. 3.143   Spring-isolation boilers.

FIG. 3.144   Point isolation of boiler.

isolated on its own structural steel frame base with four multiple steel spring mounts arranged along each side (see Figure 3.144). Each isolator is selected to carry its designated part of the total loaded weight of the boiler. Should the boiler be emptied, the resilient restraining bolts prevent the springs of the vibration isolators from overextending. Two of the restraining bolts or travel limit stops, which are normally positioned in pairs on each side of each isolator, can be seen on the isolator unit in the foreground.

Note that the gas supply piping is supported directly from the boiler base frame. The gas piping, as well as other piping connected to the boiler, is suspended on spring hangers for the length of the piping within the boiler room.

Boiler smoke outlets, or breeching connections, may be vibration-isolated by means of flexible connections having various configurations, as shown in Figure 3.146. Alternatively, manufactured flexible sections may be used (see Figure 3.146).

FIG. 3.145 **Vibration isolation for boiler.**

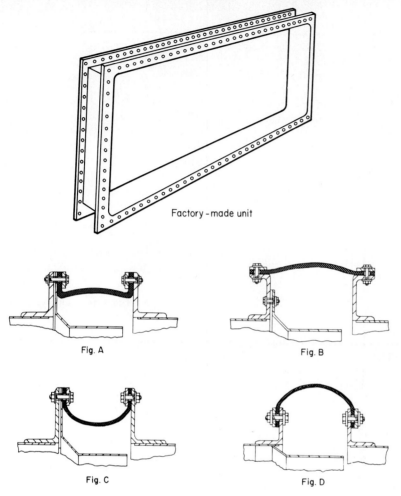

Factory - made unit

Fig. A

Fig. B

Fig. C

Fig. D

Figures A and B depict the straight sectioned composite expansion joints, the most frequently used designs.

Where extreme movement is to be accommodated, the shaped designs shown in figures C and D may be considered.

FIG. 3.146 Typical methods for boiler breeching using flexible connections (no scale). *(James Walker Mfg. Co.)*

AIR COMPRESSORS

3.147 **Air Compressor and Tank** _____

The air compressor and tank in Figure 3.147 are nicely vibration-isolated. Open, stable, steel spring isolators are properly located under cantilevered mounting brackets affixed to the concrete inertia block. In this installation, the inertia block provides a low center of gravity and helps to stabilize the rocking of the reciprocating compressor as it operates.

FIG. 3.147   **Air compressor and tank.**

3.148 **Air Compressor and Tank** _____

The air compressor and tank in Figure 3.148 are well-isolated, but not quite as neatly as the one shown in Figure 3.147. The flexible 360° loop of electrical conduit is effectively positioned. However, the isolator in the foreground is still misaligned, and each isolator, although open, has been recessed into the corners of the concrete inertia block. As a result, each unit is partly housed in a pocket of concrete. When this method is used, the pockets at the corners should be sized to maintain at least 1-in clearance between the flat top plate over the spring and the two sides of the pocket.

The mounting method shown in Figure 3.147 is preferred.

FIG. 3.148   **Air compressor and tank.**

3.149 Air Compressors

3.150 Several features of the air compressor installation in
3.151 Figures 3.149 to 3.151 are important from an acoustical
viewpoint. Figures 3.149 and 3.150 are of the same air
compressor, taken from different positions.

The compressor is mounted on its own massive
concrete base and separated from the floor slab around it
by a complete structural break (see arrow) that is filled
with resilient insulation and sealed at floor level with a
nonhardening resilient sealing compound. The entire
installation is on grade.

The electrical conduit connection seen in the foreground of Figure 3.149, although fitted with a section of
flexible tubing, was totally ineffective as a vibration-isolation device. Because of their inherent construction
features, flexible conduits of this size (1-in diameter and
larger) and length do not provide adequate vibration
isolation.

A flexible fitting similar to that shown in Figure 5.10
could be used in such installations.

In the foreground of Figure 3.150, the copper drain line
attached rigidly to the vibrating compressor and the
adjacent floor slab bridges the isolation system because
the drain line transmits vibrations from the compressor
into the floor. Although not shown, the air piping from

FIG. 3.149 Air compressor.

the compressor to the air receiver tank was isolated from the building with spring hangers, but the section of metallic corrugated bellows covered with wire braid, seen best in the foreground of Figure 3.150, does not provide the required vibration isolation. Such devices, designed primarily to compensate for expansion and contraction, are not manufactured or advertised as vibration-isolation equipment.

Regardless of this, they are relatively ineffective, but for whatever results might be obtained, they should be installed horizontally and parallel to the drive shafts of the equipment as they only flex transversely and not axially.

Figure 3.151 shows a typical detail employing several manufactured items that can be applied to achieve adequate noise and vibration control.

FIG. 3.150 **Air compressor.**

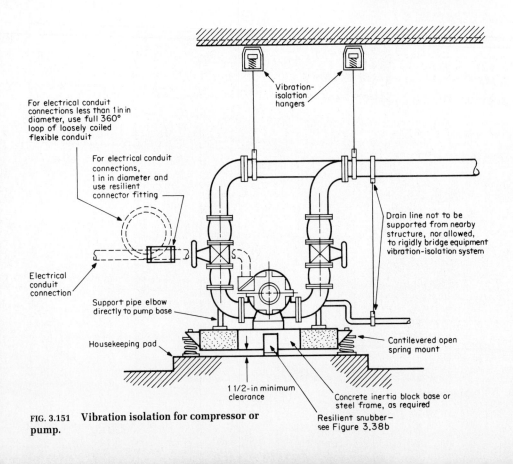

Vibration-isolation hangers

For electrical conduit connections less than 1 in in diameter, use full 360° loop of loosely coiled flexible conduit

For electrical conduit connections, 1 in in diameter and use resilient connector fitting

Electrical conduit connection

Support pipe elbow directly to pump base

Housekeeping pad

Drain line not to be supported from nearby structure, nor allowed, to rigidly bridge equipment vibration-isolation system

Cantilevered open spring mount

1 1/2-in minimum clearance

Concrete inertia block base or steel frame, as required

Resilient snubber—see Figure 3.38b

FIG. 3.151 **Vibration isolation for compressor or pump.**

COOLING TOWERS AND CHILLERS

3.152   Cooling Tower and Piping
3.153   without Vibration Isolation

The large cooling tower installation in Singapore shown in Figure 3.152 illustrates what to avoid when trying to reduce the amount of vibration entering a building.

First, although spring vibration isolators were specified for this cooling tower, none had been installed. Second, the condenser water main was partially grouted into the roof. It had no flexible connections and no vibration isolation at all.

A large meeting room directly underneath this tower was in use during the time these pictures were taken and while the installation was being operated. In this meeting room, the suspended ceiling, the light fixtures, the walls and floor, the stage, and everything in the room were vibrating, rattling, and buzzing giving a noise spectrum of about NC-55 to NC-60. The recommended acoustical environment for such a space is NC-20 to NC-30. (See Figure 1.1.)

To solve this problem, the cooling tower was put on spring vibration isolators, and all the condenser water piping to and from the tower was completely separated from the building structure and was provided with resilient isolation (see Figure 3.153).

FIG. 3.152   Cooling tower with no vibration isolation.

FIG. 3.153   Condenser water line embedded in building construction.

The multiple spring isolators in Figure 3.154 are located directly under a cooling tower. For several months, their travel limit stops were left tightly secured (see arrow). During this time, the building owner, trying to discover the cause of the excessive vibration and noise in the building, consulted the architect, the engineers, the contractors, and the equipment suppliers—all to no avail. Finally, the acoustical consultant who was retained to visit the site recommended that all travel limit stops be released and the drainpipe be detached. This pipe had been attached to the base of the steel tower support with a threaded rod pipe clamp. Because the drainpipe held down the vibration isolators rigid against the tower supports, the isolators could not function properly. As soon as the travel limit stops were released and the drainpipe was detached from the tower support base, the measured vibration in the rooms below and adjacent to the cooling tower immediately dropped to acceptable levels. A clearer understanding of the proper functioning of vibration-isolation devices could have prevented this problem.

FIG. 3.154 **Cooling tower with spring isolators.**

3.155 Cooling Tower with Spring Isolators

The cooling tower in Figure 3.155 is vibration-isolated on open, stable, steel springs with travel limit stops for controlling tower movement from wind forces. The type of isolator employed here has a three-point support system for greater spring stability. The three-point supports extend from the flat plate on top of the spring to the top steel support plate of the isolator. The top plate picks up the tower weight distributed at that point. Please note that the bottom wide flange is welded to the underside of the larger I beam (see arrow).

The special arrangement of these structural steel shapes that form the dunnage allows the isolator units to be set down inside the steel support system, thus keeping the overall height of the tower installation down to achieve as low a silhouette as possible.

FIG. 3.155   Cooling tower with spring isolation.

3.156 Cooling Tower with Spring Isolators of Variable Heights

The weight distribution points under the cooling tower shown in Figure 3.156 varied so much that vibration isolators of different heights were selected, creating an installation problem. When the isolators were delivered, the installing contractor realized that the springs had different color codings (for different weights) and solved the problem ingeniously by using a piece of standard channel iron (see arrow) of just the right height to make up the difference between the shorter isolators and the taller one. The channel iron was also as wide as the base plates of the isolators, so that it could easily be cut to the right length. The channel pieces were welded to the support base I beams, and the vibration isolators were then resiliently bolted to the channel pieces (see Figure 3.26a). The springs were adjusted to accept their load, and the travel limit stop nuts were properly released down each hold-down bolt. As a result, the installation works satisfactorily.

In spite of the contractor's ingenuity, the vibration company should have supplied isolators of equal height to avoid this field problem.

FIG. 3.156 Cooling tower with spring isolators of different heights.

3.157 Cooling Tower with Vibration Isolation

Figure 3.157 shows a Japanese-made cooling tower installation at Singapore Polytechnical University. Adequate high-frequency vibration isolation was achieved by installing two layers of 5/16-in-thick ribbed or waffle-pattern neoprene pads under the base plate of each leg of the cooling tower. The leg base plates are held in place by steel angle clips with neoprene pads cemented to the angle facing the base plate (see arrow). The pads provide the resilient separation needed between the angle clips and the base plates.

Generally, cooling towers of this size and type require spring isolators to control the transmission of unwanted low-frequency vibration.

FIG. 3.157   Cooling tower with vibration isolation.

3.158 Cooling Tower with Noise Control Enclosure

Noise control features for the cooling tower shown in Figure 3.158 included a solid concrete block wall barrier on the two exposed sides. The distance from each of the four sides of the tower to the nearest wall or barrier was recommended by the equipment manufacturer.

The tapered stack on top of the tower was added to increase the discharge air velocity somewhat, thereby helping to reduce recirculation of the warm, moisture-laden discharge air with the incoming cooler and dryer air. To reduce fan discharge noise, the stack was also treated acoustically with fiberglass duct lining on all surfaces exposed to airflow. In such applications, the duct liner should be neoprene coated, with the coated facing exposed to the airstream. This type of lining is designed to last longer in the environment of the cooling tower.

Although it was not required in this case, a considerable amount of sound-absorbing material could be added to the inside walls of the tower enclosure to absorb some sound close to the noise source.

FIG. 3.158   **Cooling tower with noise control enclosure.**

3.159 Cooling Tower with Architectural Screen

The original function of the perforated enclosure shown in Figure 3.159 is not known. However, the so-called enclosed equipment, which is on the roof of a Kansas City office building in a residential area, caused nearby neighbors to complain of the noise.

The important lesson to be learned from this installation is that any enclosure or partial housing used as an acoustical barrier must be solid — with no holes, slots, or gaps. It must have enough mass (surface weight) and be tall enough and wide enough to be capable of reducing the noise generated by the operating equipment.

The acoustical treatment recommended here was to install prefabricated metal sandwich panels, solid on one side and perforated on the other, with fiberglass between. These panels were attached to the inside of the existing equipment enclosure, thus providing a sound-absorbing noise barrier. This solved the neighbors' noise problem.

FIG. 3.159 Cooling tower without acoustical enclosure.

Figures 3.160 and 3.161 show a retrofit for a noise
problem attempted by the contractor.

The sheet-metal noise barrier for this rooftop cooling
tower was not adequate to control noise propagation for
the cooling tower. One obvious reason for the barrier's
inadequacy is the large gap between the bottom of the
barrier and the roof. Sound passes readily through this
gap. Some sound is also reflected off the roof to nearby
neighbors. Additional sound passed by the ends of the
barrier because of its insufficient length.

The fiberglass sound-absorbing treatment on the sides
of the barrier facing the cooling tower is effective in
soaking up some of the sound from the tower fan, the
waterfall noise near the tower intake louvers, and the radiated sound from the lightweight metal side panels of
the tower, which are set in motion when the tower
operates. However, unless the fiberglass is protected sufficiently against weathering effects, the degree of total
absorption at the time of installation gradually dimin-
ishes. The figures show the erosion of the fiberglass.

To provide the necessary protection, the fiberglass may
be installed in sealed polyethylene or mylar bags of
material not more than 1 mil thick; the bags will not
significantly reduce the sound-absorbing characteristics
of the fiberglass. Chicken wire placed over the entire
surface of the fiberglass will further strengthen the
installation.

An acceptable alternative treatment is to cover the
fiberglass with perforated sheet metal. Perforations
should not have diameters of less than ³⁄₃₂ in, spaced on
³⁄₁₆-in staggered centers.

FIG. 3.160   Inadequate noise control barrier for cooling tower.

FIG. 3.161   Weathering effects on unprotected sound-absorbing fiberglass.

3.162 Cooling Towers with Silencers

Some engineers and cooling tower manufacturers believe that silencers and other acoustical treatments applied to cooling towers cause problems with air movement through the tower. Figure 3.162 shows a successful installation of manufactured silencers on the discharges from two induced-draft cooling towers. The air intakes of both these towers were effectively treated with acoustically lined ducts to control noise transmitted to nearby residences, located just behind the trees in the background of the figure.

The acoustical consultant for this project worked closely with the cooling tower manufacturer's representative to design the acoustically lined ducts and silencer attachments so that they would cause the minimum additional pressure drop through the towers. The air discharge velocity from the (leaving) ends of the silencers was calculated to achieve the airspeed needed to prevent the hot, moisture-laden air from being sucked into the intake ducts. The intake duct openings were placed as far from the silencer discharge as possible while still providing the noise attenuation path required. The addition of this noise control treatment did not cause these cooling towers to have a reduced performance rating. They have functioned satisfactorily both acoustically and thermally for several years with the noise control retrofit treatments attached.

FIG. 3.162 Cooling towers with silencers.

3.163 Cooling Tower on Concert Hall Roof

Figure 3.163 shows a cooling tower in Boston on a low roof section of a concert hall. The cooling tower is close to the balcony seating, which is just on the other side of the brick wall behind the tower. To control noise transmitted from the tower, the window openings along the tower side of the building were bricked up to make them acoustically equal to the existing wall construction.

Supplemental structural steel dunnage supports the tower above the existing roof, and vibration isolators are located between the base frame of the cooling tower and the dunnage. The tower manufacturer should be consulted about whether the base frame is stiff enough to allow point isolation, as shown in Figure 3.163, without distortion of the frame or misalignment of the equipment. If the equipment base is not adequate for point isolation, new structural members must be added.

Alternatively, the vibration-isolation system may be placed beneath the steel dunnage. However, if this treatment is used, the vibration isolators must be sized to carry the weight of the dunnage frame as well as the tower. Taller isolation units might thus be needed. For aesthetic reasons, such units might require support framing to be constructed so as to keep the overall height of the installation as low as possible (see Figure 3.155).

In this installation, the entire run of supply and return condenser water piping—from the tower on the roof to the chiller and pumps in a basement mechanical room—was vibration isolated and totally separated from the building structure. In the lower right-hand portion of Figure 3.163 (see arrow) is one such piping isolator, temporarily blocked in place under the elbow of the condenser water main.

FIG. 3.163   Cooling tower on concert hall roof.

The view in Figure 3.164 is of the underside of a cooling tower installation atop a tall office building on the west coast of the United States.

It is disturbing to realize that someone selected and specified a vibration-isolation system for this tower and that shop drawings for the isolators were submitted, reviewed, and finally approved. The contractor ordered the mounts; they were shipped and then installed in positions for correct weight loading. Finally, they were adjusted and found to be adequate to perform the intended function.

Then, without care, understanding, or consideration of any sort, this huge bank of large, rigid electrical conduits was installed directly from the vibrating cooling tower to the building structure, thus bridging the vibration-isolation system and making it ineffective.

The same, well-isolated cooling tower was installed so that the top exterior framing members and the outer shell of the tower also came into direct contact with the building structure.

An installation supervisor could have spotted these obvious malpractices early in the progress of installing the equipment and could have recommended adjustments and acoustical treatment to prevent these major errors in installation.

Neatness, attention to detail, and thorough workmanship are all prerequisites to success in the installation of noise and vibration control equipment and materials.

Submittals for restrained spring isolators that are used either seismically or against wind load should include calculations showing that they will function to the proper stress limits in both planes horizontally and provide adequate protection against rocking based on the high center of gravity of the equipment.

FIG. 3.164 **Electrical conduit bank bridging cooling tower vibration isolation.**

3.165 Steam Absorption Refrigeration Machine

Because of its closeness to a concert hall that required an acoustical environment of NC-15 (see Figure 1.1), this steam absorption unit was vibration-isolated as shown in Figure 3.165. An interesting feature of the installation is that the contractor cantilevered the condensate receiver tank (see arrow 1), with duplex pumps mounted on top, off the side of the refrigeration machine's support base. Therefore, no separate vibration isolation was needed for the condensate pumps.

The two layers of neoprene pads under the spring isolators help to dampen high-frequency vibrations that may travel down the spring coils and into the base plate of the isolation unit. The figure shows how easy it is to check alignment and spring deflection of an unhoused isolation unit.

A common hazard of this type of installation occurs when support bases for vibrating equipment are installed with inadequate clearance underneath. In Figure 3.165, to the right of the spring isolator, a short section of pipe appears, under the support base (see arrow 2). Pieces of construction debris often get under these bases and lodge beneath them, negating the effectiveness of the vibration-isolation system. A minimum of 1 in of clearance should be maintained under all equipment bases. After work is completed, an essential feature of job supervision is to check the space under the equipment bases to see that it is completely free of obstacles.

FIG. 3.165 Steam absorption refrigeration machine.

3.166 Important Details of Installation

Several small but extremely important features may be noted in the installation shown in Figure 3.166. First, all construction debris, i.e., pipe nipples or short sections of pipe (see arrow), pieces of wood or concrete, cans, bottles, nuts, bolts, and the like must be removed from under steel frames or concrete inertia blocks supporting vibrating equipment; otherwise, such items can seriously reduce the effectiveness of the entire vibration-isolation system for that piece of equipment. Please look closely and you will note examples of the above in the figure. Extra special care and attention will be required to see that this does not happen on your project, for it is a very common occurrence.

Second, the copper drain line is installed correctly except that it contacts the spring coil of the isolator on its way to the floor drain funnel. More careful planning and forethought should be used to preclude such situations.

Note that this drain pipe is so close to the floor drain that no intermediate supports were used. In cases where it is necessary to run drains extended distances to the nearest floor drain, intermediate supports may be required. Adequate vibration isolation between the drainpipe and building floor must be provided. Usually this can be accomplished with the effective use of neoprene isolators or pads or devices similar to those shown in Figure 4.8.

FIG. 3.166 Inertia base and obstructions.

FANS—SUSPENDED AND FLOOR MOUNTED

3.167 Fan with Inertia Base and Spring Isolators

On this project in Sydney, Australia, the fan and its drive motor have been rigidly mounted on a steel frame concrete inertia block. The block is isolated at several points around its edge on open, stable, steel springs sitting in neoprene cups. This installation is properly installed and will work well.

The entire assembly is on a housekeeping pad about 4 in high. Such pads, although not generally recommended for acoustical purposes, are very effective in protecting vibration-isolation units from rust and corrosion when mechanical equipment rooms are washed down.

Fans like the one shown in Figure 3.167 operate at total system static pressures of 2-in water gauge or higher, air thrust developed by the fan should be checked against the horizontal spring stiffness. When the thrust of the fan exceeds 10 percent of the total horizontal spring stiffness, the concrete inertia base should be sized to increase the total weight of the system until the fan thrust is 10 percent of the new horizontal stiffness or less. This procedure will eliminate the need for snubbers which may transmit vibration.

FIG. 3.167 Fan with inertia base and spring isolators.

3.168 Resilient Snubbing Device

An actual resilient snubbing device similar to that shown in Figure 3.38b is seen in Figure 3.168. Manufacturers of vibration-isolation devices will generally supply such equipment along with their vibration isolators.

However, the contractor needs to do some coordinating with the mechanical equipment supplier so that snubber height and angle leg lengths will suit the particular application in order for adequate snubbing of motion of the particular equipment to be achieved.

FIG. 3.168 **Resilient snubbing device.**

3.169 Improper Installation of Hanger-Type Isolator

Figure 3.169 is one of four identically installed vibration isolators for a ceiling-hung axial fan in New Hampshire. The installation has two problems.

First, the isolator is installed upside down, with the spring on top and the neoprene element on the bottom.

The neoprene element is stiffer than the steel spring and its effectiveness depends on the resistence behind it. With the neoprene element on top, the element is working directly against the rod connected to the structure. With the neoprene element on the bottom, in the incorrect location, the vibration comes from the equipment, pulls on the neoprene, which in turn pulls on the hanger box and then against the spring element, which is extremely resilient. Thus, the neoprene element is far less effective in this position.

Second, the hanger box is resting directly against the fan housing. This position makes the isolator useless.

In installations like this, isolators should be correctly positioned according to the manufacturer's installation instructions (right side up!). Each isolator should be located in its rod at a height that will permit easy inspection and adjustment and will allow the hanger to rotate a full 360° horizontally without touching the equipment that it is isolating, the building construction, or any nearby systems. Hanger boxes do rotate during system operation; therefore, it is essential to leave space for full rotation when installing these units.

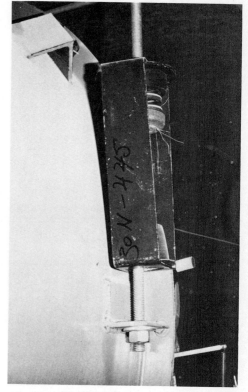

FIG. 3.169 Upside-down short-circuited vibration-isolation hanger.

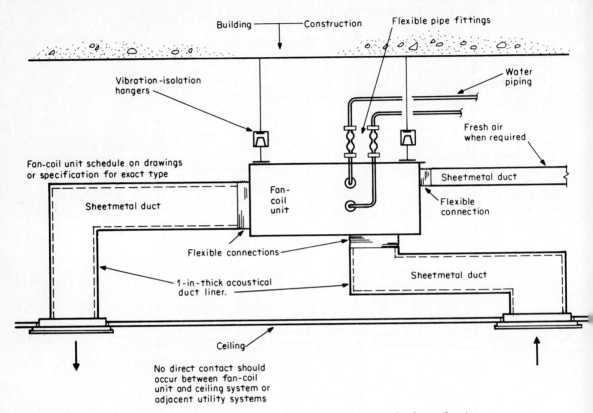

FIG. 3.170   **Typical noise and vibration control treatment recommendations for fan coil units.**

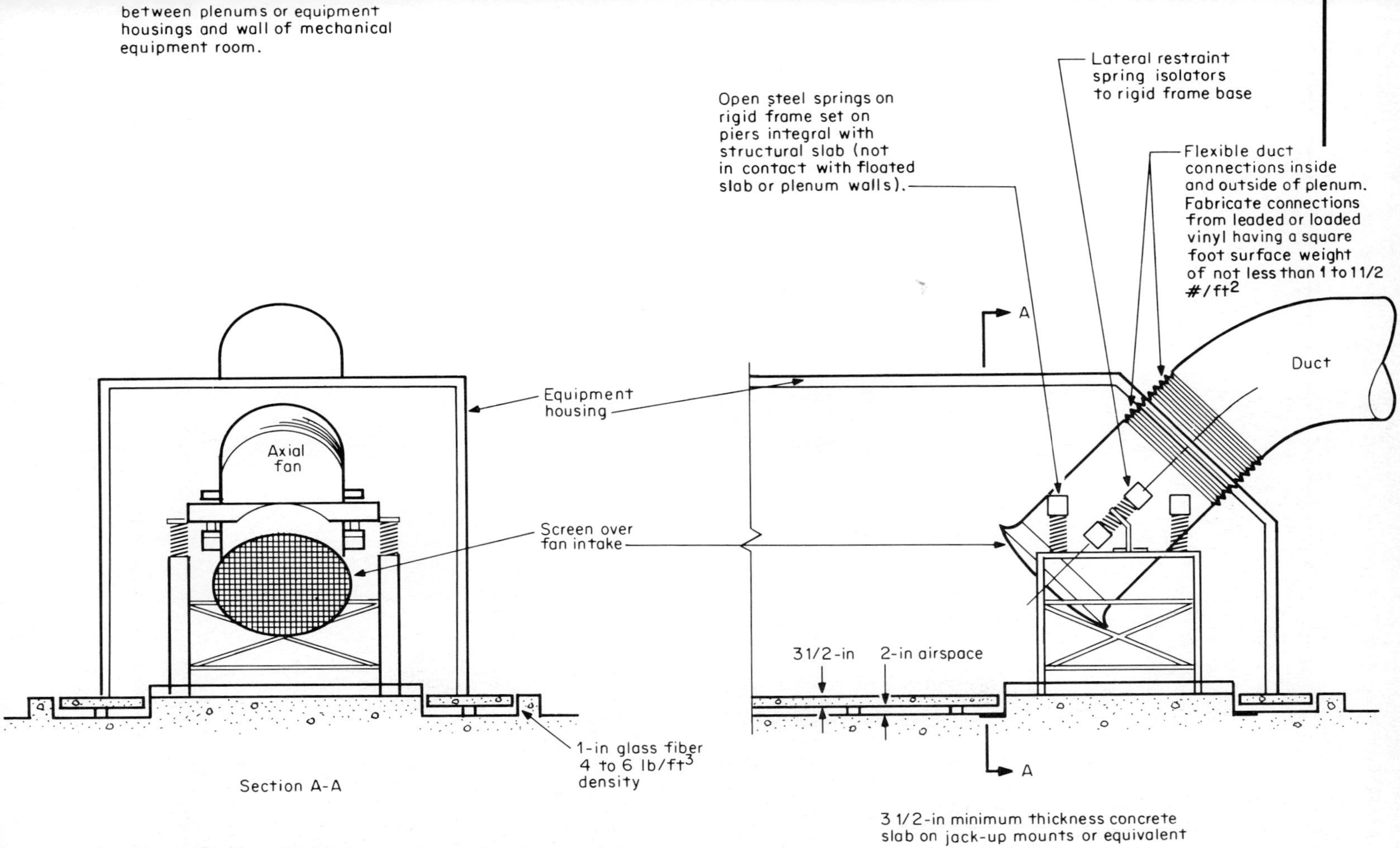

between plenums or equipment
housings and wall of mechanical
equipment room.

Lateral restraint
spring isolators
to rigid frame base

Flexible duct
connections inside
and outside of plenum.
Fabricate connections
from leaded or loaded
vinyl having a square
foot surface weight
of not less than 1 to 11/2
#/ft2

Duct

Open steel springs on
rigid frame set on
piers integral with
structural slab (not
in contact with floated
slab or plenum walls).

A

A

3 1/2-in minimum thickness concrete
slab on jack-up mounts or equivalent

3 1/2-in   2-in airspace

Section A-A

Equipment
housing

Screen over
fan intake

1-in glass fiber
4 to 6 lb/ft3
density

Axial
fan

FIG. 3.171   Vibration isolation for axial housed fan application.

AIR-HANDLING UNITS

3.172 Correct Vibration Isolation of Air-Handling Unit

The large unit shown in Figure 3.172, installed in Montreal, Quebec, has been mounted on open, stable, steel spring isolators. Note the excellent alignment of each spring. The figure shows that when these springs are specified to have a horizontal stiffness approximately equal to their vertical stiffness, they will remain in correct alignment.

The steel angle brackets, which are cantilevered off a concrete inertia block, provide good clearance all around the spring for ease of movement and inspection. In the right foreground of the figure is a neoprene snubbing device, installed to restrict the movement of the mounted equipment (see arrow).

Snubbing devices are not generally recommended except to control unusual movement problems. Their use should be avoided where control is needed under steady operation.

FIG. 3.172   Correct vibration isolation of air-handling unit.

3.173 Incomplete Installation of Vibration Isolators under Air-Handling Unit

Figure 3.173 vividly illustrates an incorrect installation. The isolators shown are the correct type for isolating rooftop mechanical equipment. The vertical rods, welded to the isolator base plate and extending upward through an oversized hole in the isolator top plate, act as travel limit stops. They limit the degree of equipment movement when wind forces occur.

However, to perform their function, these isolators must be firmly secured to their supporting base. As the figure shows, at least three of the units are not even touching the supporting base, leaving the air-handling unit supported by the ducts, pipes, and conduits connected to it.

The recommended method for holding down this type of vibration unit is shown in Figure 3.33.

FIG. 3.173   Incomplete installation of vibration isolators under air-handling unit.

3.176 In these views of the air-handling unit seen in Figure 3.174, the vibration-isolation treatment has been corrected. Figure 3.174 shows the whole air-handling unit. Figure 3.175 shows the base of the isolator firmly secured to the concrete base. Neoprene washers, installed between the steel washer and the steel base plate of the isolator, provide the required resilient separation.

Two layers of waffle-pattern neoprene pads have been installed under the entire base plate of the isolator. A single neoprene pad is standard equipment for this isolator. The single pad (see arrow) goes under the smaller steel base plate that supports the two springs between the two vertical steel travel limit rods. Not all isolators come with this neoprene pad. Thus, the two additional layers of neoprene are recommended to ensure sufficient damping of high-frequency vibrations traveling down the spring coil. A single layer of neoprene is frequently rendered ineffective by construction debris, such as cement grout or plaster, but a two-layer thickness is not so easily affected.

Evidence of rust on this installation in Figure 3.175 shows the need for specifying that outdoor installations of vibration isolators must be cadmium-plated or neoprene-coated for adequate protection and durability.

Figure 3.176 shows that the travel limit stop nut, just under the isolator top plate, has been released. This leaves enough free space between the nut and the top plate to permit the spring to function properly. When subjected to wind-load effects, the air-handling unit will move only slightly. Further movement will be restricted by the travel limit stop nut.

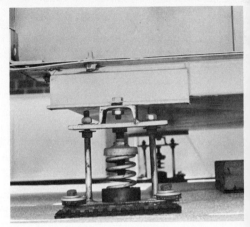

FIG. 3.174 **Rooftop air-conditioning unit.**

FIG. 3.175 Vibration isolators properly secured.

FIG. 3.176 Vibration isolators secured to base.

3.177 Air-Handling Unit with Overloaded Isolation Pads

Figure 3.177 is an example of an overloaded vibration-isolation pad treatment. The excessive weight of the air-handling equipment compresses the neoprene cork pads so much that they lose their resiliency and effectiveness as a vibration isolator. All vibration-isolation equipment and materials should be properly checked for weight loading so that the manufacturer's recommendations are not exceeded.

In this example, the pads were probably selected for the proper weight without any attention being paid to the size of the leg resting on them. Had the pads been furnished with a steel plate on top, covering the pads completely, this problem probably could have been avoided.

FIG. 3.177 Air-handling unit with overloaded isolation pads.

3.178 **Freestanding Enclosure for Suspended**
3.179 **Air-Handling Unit** ─────────────────────────────

3.180 Figures 3.178 to 3.181 are before-and-after pictures of
3.181 noise control treatments in a large, open plan school. The
casing-radiated noise levels from this 20,000 ft³/min
air-handling unit were so high that when the building
was first used, effective teaching was impossible in a
large percentage of the assigned teaching areas in this
open plan space. The air-handling unit was located on a
suspended platform, shown in Figure 3.178, just below
the underside of the roof of a huge, open, multilevel
teaching and study space. The original design team did
not include a consultant for noise and vibration controls.
To solve the noise problem, an acoustical consultant was
engaged to conduct sound-level measurements, perform
an analysis of the noise control required for an acceptable acoustical environment at all teaching areas, and
recommend noise control treatments. The consultant
provided detailed drawings, specifications, and installation techniques for appropriate equipment and materials.

Figure 3.180 shows the noise control treatment, which
was acoustically, structurally, and architecturally
successful. The predicted noise reduction was achieved,
allowing all teaching and study areas to be used for their
appropriate activities.

A key factor in the success of this effort was the close
supervision of the installation at scheduled intervals.

FIG. 3.178　Air-handling unit in open plan school with no control treatment.

This air-handling unit was the largest of many such units; it therefore produced the highest noise levels when operating. The noise control treatment consisted mainly of a complete freestanding enclosure made of wood studs and ⅝-in plywood panels, sealed airtight. A 4-in-thick blanket of fiberglass insulation was applied between the studs, so that all the walls and the roof of the enclosure were covered. The enclosure itself and all penetrations of it by ducts, pipes, and conduits were acoustically sealed airtight. One fully gasketed access door was provided, and the enclosure was designed and placed on the supporting structure so that a person inside it would be able to walk around all four sides of the airtight equipment for maintenance and servicing. For adequate visibility, some lighting was added inside the enclosure. Special care was taken to keep the air-handling equipment and the connecting pipes, ducts, and conduits from touching the freestanding enclosure.

The ⅝-in-thick plywood was selected to provide the required noise transmission loss, and the fiberglass inside the enclosure was applied to control the interior noise buildup by absorbing much of the noise very close to its source. The enclosure reduced noise levels in teaching areas from noise criteria (NC) level NC-60 to level NC-45. Figure 3.180 shows the noise reduction predicted, and Figure 3.181 shows the actual noise reduction. A certain amount of background sound was necessary in this space

FIG. 3.179   **Air-handling unit in open plan school with noise control treatment installed.**

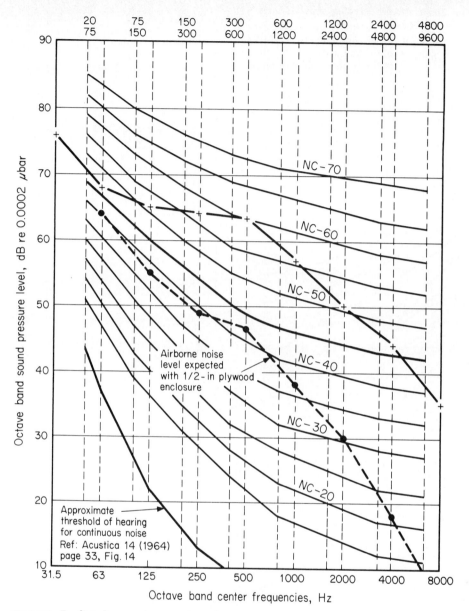

Top axis values:
| 20 | 75 | 150 | 300 | 600 | 1200 | 2400 | 4800 |
| 75 | 150 | 300 | 600 | 1200 | 2400 | 4800 | 9600 |

NC-70

NC-60

NC-50

Airborne noise
level expected
with 1/2-in plywood
enclosure

NC-40

NC-30

NC-20

Approximate
threshold of hearing
for continuous noise
Ref: Acustica 14 (1964)
page 33, Fig. 14

Octave band sound pressure level, dB re 0.0002 μbar

Octave band center frequencies, Hz

FIG. 3.180 Predicted approximate noise reduction with acoustical treatment.

to prevent the voices in one teaching area from being
heard in another area. The level of NC-45 provided the
required level of background sound to prevent speech
interference between teaching groups.

Manufactured duct silencers were installed on both
supply and return air ducts of the air-handling unit to
control fan noise transmitted down the ducts by the

supply and return air fans. These silencers were strategically placed so that sections of ducts located between the fans and their silencers were kept within the acoustical enclosure.

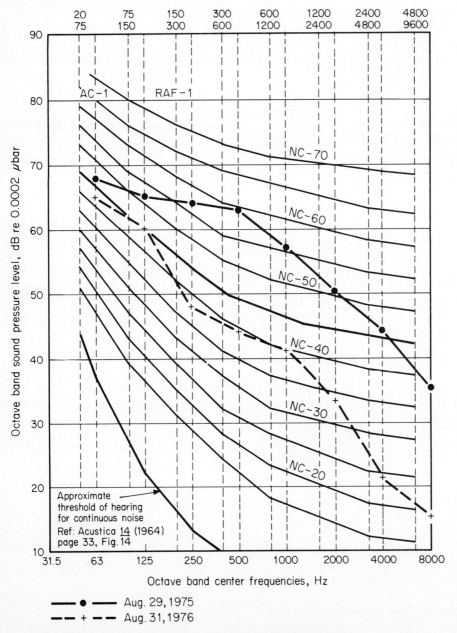

FIG. 3.181   **Actual noise reduction achieved with acoustical treatment.**

3.182 Acoustical Enclosure for
3.183 Air-Handling Unit

3.184
3.185
3.186
3.187

The air-handling units shown in Figure 3.182 are typical of several such units installed in an open plan school. These air-handling units vary in size from 4000 to 7000 ft³/min. The supply air for the units is distributed through sheet-metal ducts. In most cases, the return air is drawn directly from the space served, without duct-work, as shown in Figure 3.182. The noise produced by these units was so great as to require the noise control treatment shown in Figures 3.182, 3.183, and 3.184.

The acoustical enclosure shown in Figure 3.183 is for reducing noise levels generated by the air-handling unit electric motor and belt drives. The enclosure is made of 20-gauge galvanized sheet metal, and all its interior surfaces are lined acoustically with a glass fiber insulation with black neoprene-coated facing exposed to the interior. The hole cut in the sheet-metal casing, which can be seen at the bottom of Figure 3.183, is for a duct connection that will permit entry of a small amount of conditioned air from the main supply trunk duct, for ventilating the enclosure. Air escapes from the enclosure through a small, acoustically lined elbow vent. The enclosure may be seen in place on a typical air-handling unit in Figure 3.184.

Figure 3.184 also shows the air-handling unit, covered with two 2-in-thick layers of glass fiber insulation applied over the entire exterior of the unit. Certain sections of the insulation are removable, for access to or servicing of the air-handling unit components. Figure 3.184 also shows a conical duct silencer that has been added to reduce fan noise.

FIG. 3.182 Acoustically untreated air-handling units.

Figure 3.185 shows the last step of noise control treatment, an outer skin of 1 lb/ft² lead.* The whole acoustical treatment is held in place on the air-handling unit by means of weld nail pins (see Figure 3.186). Their circular heads are clearly visible in the figure. Weld nail pins are the preferred attachments for this type of enclosure. The weld-nail-pin attachment is the process of welding a stud or pin to a metal plate or other metal surface through the instantaneous discharge of stored electrical energy. This system is also known as *capacitor-discharge* or *stored-energy* welding. It combines the advantages of resistance and arc welding but eliminates the possibility of burn-through or distortion even when welding to light-gauge metal.

The energy stored in banks of low-voltage dc capacitors is released through the end of the stud or pin by triggering the gun. The high concentration of heat, produced by a current of about 300,000 A/in², results in an arc capable of welding the stud or pin to the work piece. The complete cycle takes only 4 ms. Stud welding is a positive, quick, and dependable means of fastening. No special preparation or skills are needed. The only utility required is 115 V ac.

Removable sections of the lead and fiberglass treatment are secured and sealed by Velcro strips that run around the entire perimeter of each removable section. The overlaps and joints of the lead sheets were sealed airtight with a special lead adhesive.

This noise control treatment, installed with first-class workmanship, performed as expected within the 2- or 3-dB allowance. An alternative external acoustical treatment for noisy air-handling equipment is shown in Figure 3.187.

This is another example of how attention to details and close job supervision can achieve pleasing and functional acoustical environments.

Note that the original vibration isolators for the air-handling units continued to perform satisfactorily, even when the added weight of the noise control treatment was applied.

FIG. 3.183   **Air-handling unit motor and U-belt drive acoustical housing.**

---

* Mass-loaded vinyl of the same surface weight of the lead is an acceptable alternative. See tabulation below, which is based on a surface weight of 1 lb/ft² with a sound transmission class of 27, for noise-transmission-loss data, which are essentially the same for both materials.

Center frequency, Hz	125	250	500	1000	2000	4000
Sound pressure level, dB	16	18	23	28	33	38

FIG. 3.184   Air-handling unit with fiberglass attached to exterior.

FIG. 3.185   Air-handling unit with lead sheet applied over glass fiber.

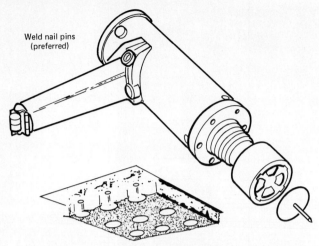

Weld nail pins
(preferred)

It is the process of welding a stud or pin to a metal plate or other metal surface through the instantaneous discharge of stored electrical energy. This system is also known as capacitor-discharge or stored-energy welding.

It combines the advantages of resistance and arc welding but eliminates the possibility of burnthrough or distortion even when welding to light-gauge metal.

The energy stored in banks of low-voltage d.c. capacitors is released through the end of the stud or pin by triggering the gun. The high concentration of heat, produced by a current of about 300,000 A/in^2, results in an arc capable of welding the stud or pin to the work piece. The complete cycle takes only 4 ms. Stud welding is a positive, quick and dependable means of fastening. No special preparation or skills are needed. 115V. a.c. is the only utility required.

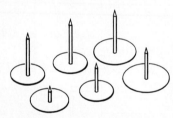

Stic-klips
(alternative)

These cupped-head pins will secure all types of insulating materials including fiberglass, polyurethane, or other foam, etc. There is no need to drill or locate holes or punching. Available in a variety of lengths and diameters.

Type N klip for attaching a variety of cellular insulation materials, such as fiberglass, urethane, styrofoam, etc., to ceiling and wall surfaces.

Type A klip for attaching various types of low-density insulation materials.

Type B klip for attaching medium-density insulation materials.

FIG. 3.186   **Attachments for acoustical enclosures.** *(AGM Industries, Inc.)*

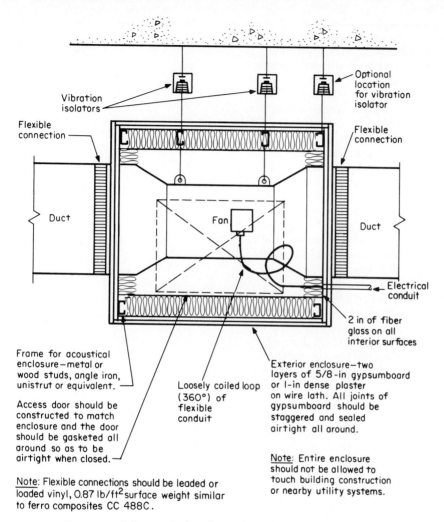

Vibration
isolators

Optional
location
for vibration
isolator

Flexible
connection

Flexible
connection

Duct

Duct

Fan

Electrical
conduit

2 in of fiber
glass on all
interior surfaces

Frame for acoustical
enclosure—metal or
wood studs, angle iron,
unistrut or equivalent.

Loosely coiled loop
(360°) of
flexible
conduit

Exterior enclosure—two
layers of 5/8-in gypsumboard
or 1-in dense plaster
on wire lath. All joints of
gypsumboard should be
staggered and sealed
airtight all around.

Access door should be
constructed to match
enclosure and the door
should be gasketed all
around so as to be
airtight when closed.

Note: Entire enclosure
should not be allowed to
touch building construction
or nearby utility systems.

Note: Flexible connections should be leaded or
loaded vinyl, 0.87 lb/ft² surface weight similar
to ferro composites CC 488C.

FIG. 3.187 **Recommended acoustical enclosure for in-line suspended fans.**

3.188 Acoustical Treatment of
3.189 Air-Handling Units

3.190 Figure 3.188 shows two air-handling units (AHU) located in a shop, prior to any external acoustical treatment.

Figure 3.190 is of the same units, with sound-isolating boxes installed over the motor and belt drives of each of the two air handlers. The acoustical treatment seen in these pictures was limited to the motors and belt drives, because they had been identified, during a series of noise measurements conducted at the jobsite, as the dominating noise sources. Although the motor and drives were not the sole sources of noise, the noise contributed by these components was significantly in excess of the other noise source, so that a substantial noise reduction could be achieved by treating only these components.

Sometimes the noise reduction achieved by acoustically treating the motor and belt drive of an air-handling unit is sufficient. Such was the case in this high school shop area, which is typical of several such shops used for instructing students studying electricity, plumbing, mechanical equipment, and sheet-metal working. Although such spaces are not considered acoustically sensitive, when teaching and conveying explicit instructions to the students is vital, noise levels in the speech intelligibility range (500 to 4000 Hz) must be controlled to levels that will allow good understanding of word communication.

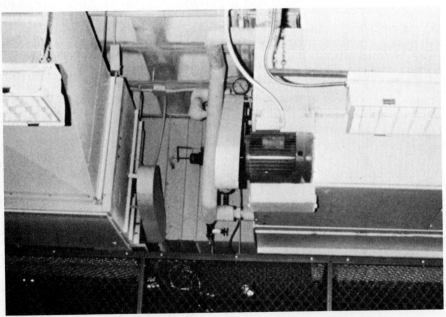

FIG. 3.188   Suspended air-handling unit with no acoustical treatment applied.

The operating noise levels produced by the acoustically untreated air-handling units were such that effective teaching at any location within a given shop area was impossible. The sound-absorptive fiberglass panels installed in the pan areas of the concrete ceiling offer some degree of improved room acoustics. However, shop areas are generally of all hard material—concrete ceiling, concrete floor, and concrete block walls—and are thus highly reverberant. Even equipment in the shops is hard, in that it is generally made of steel and iron. Therefore, sound generated in such a space reflects from all these hard surfaces many times with very little decay or energy loss, and the resulting sound levels tend to build up and linger, making good speaking and listening conditions extremely difficult.

Figure 3.189 shows a close-up view of one of the AHU motor enclosures and its access door and ventilation air-duct connections.

FIG. 3.189 Air-handling unit with motor and drive acoustical enclosure.

FIG. 3.190 Air-handling unit with motor and drive acoustical enclosure.

3.192 Acoustical enclosures for noisy mechanical equipment may also be made from either structural steel shapes, metal studs, or special manufactured shapes like Unistrut. Over this framework an exterior skin of plywood or gypsum board may be secured and sealed airtight. In Figures 3.191 and 3.192, the exterior is fabricated from gypsum board with steel corner bead to protect the edges. Figure 3.191 shows how the access door of the enclosure should be framed. This method allows for gaskets and cam locks so that the door can be made tight. The equipment being enclosed in these figures is an in-line axial fan for exhaust systems in toilet areas. The fan is placed quite close to a concert hall; therefore, the extra noise control treatment was required.

To save time and money, the contractor elected to make the fan enclosures three-sided and to use the slab above as the fourth side of the enclosure box (see arrow). This procedure was not what was recommended. It is important to keep the enclosure separate from the building, so that neither structural nor airborne excitation can travel into the building construction.

Because noise from both these fans was audible in the concert hall, it became necessary to cut away the enclosures from the slab and building construction and install a separate top and also to make sure that the fan and its enclosure were completely vibration-isolated on spring hangers. Since these corrective measures have been implemented, the fan now operates without being heard in the concert hall.

FIG. 3.192 **In-line fan acoustical enclosure.**

FIG. 3.191 **In-line fan acoustical enclosure.**

3.194 Sometimes enclosures are required to control noise
entering a space from outside. Such a case is shown in
Figures 3.193 and 3.194.

Figure 3.193 shows an exhaust fan located on the roof
of a stage house serving a concert hall in Puerto Rico. It is
typical of such exhaust fans.

The exterior acoustical treatment of the duct coming
through the stage roof to the exhaust fan inlet controls

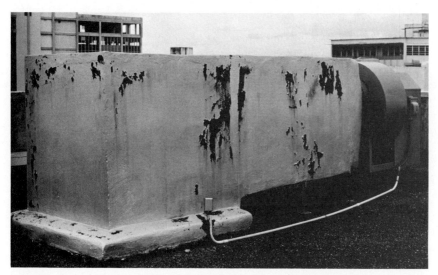

FIG. 3.193 Stage house roof exhaust fan with exterior acoustical treatment.

FIG. 3.194 Acoustically treated ducts at underside of stage house roof.

city activity noise from cars, trucks, sirens, and aircraft flyovers from entering the stage area.

The treatment in these instances was built up from angle irons and gypsum board with an exterior waterproof coating over the entire installation, which was then painted. The paint can be seen coming off the waterproof treatment. Special bonding coating had to be used later to keep paint on the enclosure. The fans are mounted on open, stable, steel spring vibration isolators.

To control the fan noise so that it would not be annoying to persons performing on stage, acoustically treated intake ducts were designed for installation at the underside of the stage roof. Careful coordination is advised in such applications, so that interferences with stage rigging are kept to an absolute minimum. Figure 3.194 illustrates the importance of this recommendation.

UNIT HEATERS

3.195 Unit Heater Vibration Isolation

Where required or recommended, unit heaters like the one pictured, cabinet heaters, and fan coil units should be vibration-isolated as shown in Figure 3.195. The piping connected to such equipment for a distance of not less than 50 ft from the unit may also require isolation.

FIG. 3.195   Unit heater vibration isolation.

ROOFTOP EQUIPMENT

3.196 Rooftop Air-Conditioning Unit ─────────────────────

3.197 The large, self-contained air-conditioning unit shown in
3.198 Figure 3.196 was installed on the roof of a hospital in
3.199 Maine. When it was put into operation, several nearby
neighbors complained about the noise of the equipment,
which was particularly annoying at night.

Figure 3.197 shows the relationship between the
source of the noise (the air-conditioning unit), the noise
path, and the receiver of the noise (a private residence).
The figure also shows the back of the partial acoustical
enclosures (barriers) that were designed and installed to
reduce noise levels at the hospital property line to a
maximum of 50 dBA at night.

Figure 3.198 shows a close-up of the acoustical treatment, consisting of sound-absorbing manufactured

FIG. 3.196 **Rooftop equipment prior to application of any noise
control treatment.**

FIG. 3.197 **Equipment with noise control barriers in place (back
side of barriers shown).**

FIG. 3.198 Close-up of back side of barriers.

FIG. 3.199 Front side of installed noise control barriers.

panels installed at roof level and on top of the air-conditioning unit. The roof panels were needed to reduce the noise of the fans and compressors at the sides and bottom of the unit. The panels on top of the unit were needed to control noise transmitted from the two large propeller condenser fans. Before installation of the top panels, the existing spring vibration isolators were checked to make sure that they could support the added weight of the panels. The location, size, and shape of the panels were specified to conform to the equipment manufacturer's recommendations for ensuring proper airflow in and out of the equipment components.

Figure 3.199 shows a view of the exterior of the barriers, painted and well-designed to give them a pleasing appearance.

3.200 Rooftop Equipment on
3.201 Vibration-Isolation Curb

3.202 Figures 3.200 and 3.201 show a typical installation of rooftop equipment mounted on a vibration-isolation curb. Details of this curb are illustrated in Figure 3.202. The vibration-isolation curb is usually fabricated from extruded aluminum shapes with a top member overlapping the bottom member. Corners of the curb mount can be made tight by welding. Springs can provide up to 1½ in deflection. Resistance to wind loads is achieved by means of internal resilient snubbers. The resilient weather seal between the top and bottom members is usually a neoprene strip or closed-cell sponge rubber material. When properly installed, the curb looks neat and finished.

Two vibration paths that were overlooked in Figure 3.201 are the electric conduit and the gas-pipe connections. These two lines were both rigidly connected to the rooftop unit, thus providing paths along which vibrations may travel to the building.

FIG. 3.200   **Exterior of vibration-isolation curb.**

FIG. 3.201   **Air-conditioning unit on vibration-isolation curb.**

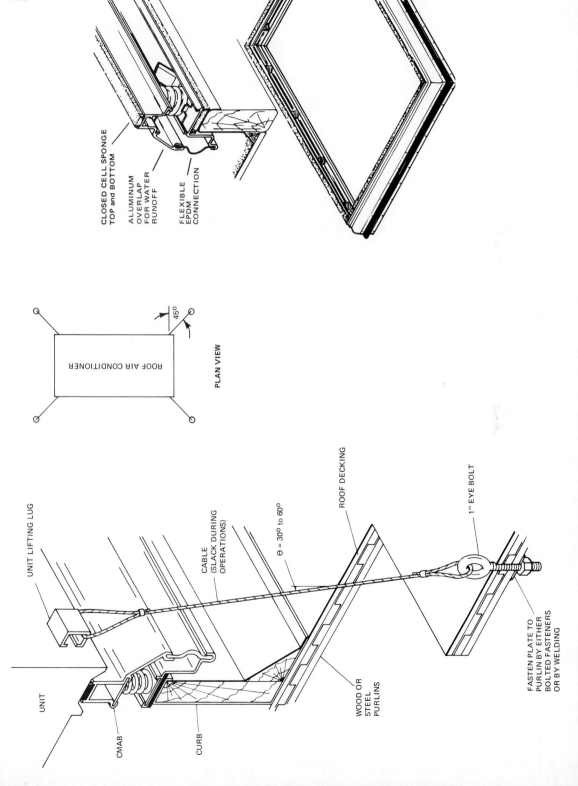

CLOSED CELL SPONGE
TOP and BOTTOM

ALUMINUM
OVERLAP
FOR WATER
RUNOFF

FLEXIBLE
EPDM
CONNECTION

ROOF AIR CONDITIONER

PLAN VIEW

45°

UNIT LIFTING LUG

CABLE
(SLACK DURING
OPERATIONS)

Θ = 30° to 60°

ROOF DECKING

1" EYE BOLT

UNIT

CMAB

CURB

WOOD OR
STEEL
PURLINS

FASTEN PLATE TO
PURLIN BY EITHER
BOLTED FASTENERS
OR BY WELDING

FIG. 3.202   Spring-isolated curb for rooftop-mounted equipment. (*Mason Industries.*)

3.205 When they are erected, acoustical panel barriers, whether factory-fabricated or field-constructed, should be structurally supported and braced adequately. Particular thought should be given to structural design features that will sufficiently resist anticipated wind-load effects. Two barrier installations are shown, one in Figures 3.204 and 3.205 and the other in Figure 3.203.

In both cases, the panels were permitted to rest on the roof deck with a vinyl closer strip along the bottom to close off the bottom of the panels from the roofing.

These panel installations both employ "sandwich" construction, made up of a solid sheet-metal outer skin with a thickness of fiberglass and a perforated sheet-metal inner skin. The solid outer skin of the panel provides the noise-transmission-loss capability. Sound absorption is provided by the perforated inner skin opening to fiberglass fill. Also available are panels that have perforated outer skins on both sides, with 2-in sound-absorbing material behind each perforated face and with a solid septum panel in the middle. In certain applications, this type of panel can be quite effective for both noise transmission reduction and sound absorption on both sides.

The exterior of the panel barrier seen in Figure 3.204 may be viewed in Figure 3.205. The appearance is quite neat and clean, and the panels may be painted if so desired.

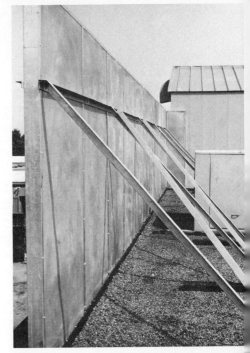

FIG. 3.204   **Back side of noise control panels and supports.**

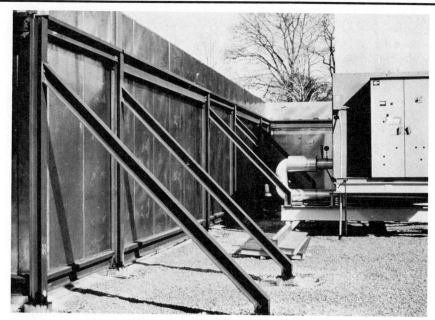

FIG. 3.203   Acoustical panel barrier for rooftop equipment.

FIG. 3.205   Front side of acoustical barrier panels.

3.206 Rooftop Air-Cooled Condensing Units with Acoustical Panels

Even small air-cooled condensing units like those shown in Figure 3.206 can create noise problems when private residences are located nearby. In this case, residential neighbors are behind the row of trees, which incidentally, do nothing to reduce the noise.

The most convenient and practical solution here was to design a barrier that would provide the noise control required to achieve a 50-dBA nighttime noise level at the owner's property line. These manufactured panels, acoustically absorptive on one side only, worked well. A structural engineer designed supports for the panels to offset the anticipated effects of wind loading. The panels were sealed along the roof joint and where they abutted the penthouse to the right.

FIG. 3.206 **Rooftop air-cooled condensing units with acoustical barrier panels.**

CONCRETE INERTIA BLOCKS

3.207 Concrete Inertia Blocks
3.208 for Floor-Mounted Equipment

3.209 Manufactured steel frames with reinforcing rods for
3.210 concrete inertia blocks are seen in Figure 3.207. The
3.211 frames are ready to receive concrete and have been
3.212 placed on top of polyethylene sheeting material to keep
3.213 the concrete to be poured into the frames separate from
3.214 the existing concrete floor.

Figures 3.208 and 3.209 show progressively how the
concrete is poured and screeded. Figure 3.210 illustrates
the finished inertia blocks with the equipment hold-
down bolts positioned and embedded in the newly
poured concrete. Note that steel cantilevered brackets
are welded to the steel frame as an integral part of it.

Unless special conditions dictate a structural analysis
for the concrete inertia block, a general rule of thumb
currently applied for block thickness is one-twelfth the
longest dimension of the equipment supported, and in no
case less than 6 in thick. Concrete blocks need not be
thicker than 12 in unless specifically recommended.

FIG. 3.207   Inertia base frames ready to receive concrete pour.

FIG. 3.208   Concrete being poured into inertia base frames.

FIG. 3.209   Concrete being leveled off with top of inertia base frame.

The floor plan shape of inertia blocks should be configured to suit each piece of equipment mounted on the block. This is particularly true in case of pumps where it is desired to pick up the weight of connected piping on the inertia blocks. Figure 3.212 shows the inertia base configured in T shape for a horizontally split double suction pump. Both intake and discharge pipe elbows are supported rigidly to the inertia base frame. Figure 3.213 is for an end suction pump, which requires a converted rectangular-shaped base with only the intake pipe elbow requiring rigid support to the base frame.

Once a piece of vibrating equipment has been properly isolated, care should continue to be exercised to see that utility connections to the equipment do not bridge or bypass the isolation system. Figure 3.211 illustrates one such case in point where this detail was overlooked. Note how the copper city water piping (see arrow) for system makeup has been rigidly connected to the pumps and then to housekeeping pads under the isolated pumps. This arrangement permits unwanted vibration to transmit to the building. The copper pipe has been connected to the housekeeping pads with standard rigid split pipe clamps.

Drainpiping from the pump base to the equipment room floor drain can be another path by which vibration travels into the building construction. See Figure 3.214.

Both city water piping and drainpiping can be vibration-isolated satisfactorily by using soft (30-durometer) neoprene pads under pipe supports or wraps of neoprene strips between the pipe and pipe clamps.

FIG. 3.211 **Inertia block isolation-bridged by city water piping.**

FIG. 3.210 **Poured inertia base being cured.**

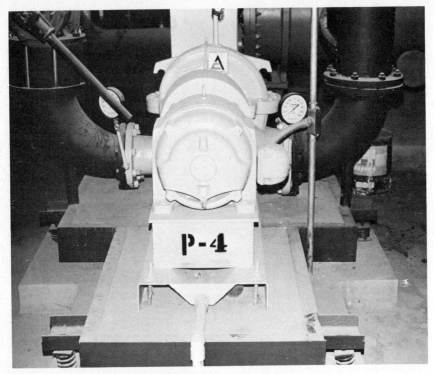

FIG. 3.212   T-shaped inertia block for double suction pump.

FIG. 3.213   Rectangular-shaped inertia block for end suction pump.

The shape of concrete inertia blocks should be fabricated to accommodate the suction and discharge pipe elbows from different type pumps. Note the T-shaped block in Figure 3.212 and the rectangular shape in Figures 3.213 and 3.214.

FIG. 3.214   Inertia block isolation-bridged by drain piping.

Plumbing Pipes

4

4.1 Vertical Plumbing Lines

4.2 The copper pipes in Figure 4.1 are plumbing lines
4.3 serving toilets. They run up through several floors of a
4.4 multilevel building in Caracas, Venezuela.

FIG. 4.1 Vertical plumbing lines.

Minimum of two layers of 5/16-in-thick ribbed- or waffle-pattern neoprene to provide separation between pipe clamp and building.

Conventional pipe clamp

10 gauge steel bearing plate

Floor construction

16 gauge separator plate

Pipe

Metal sleeve provided by mechanical contractor

Pack with fibrous material all around

Seal top and bottom with nonhardening resilient sealant

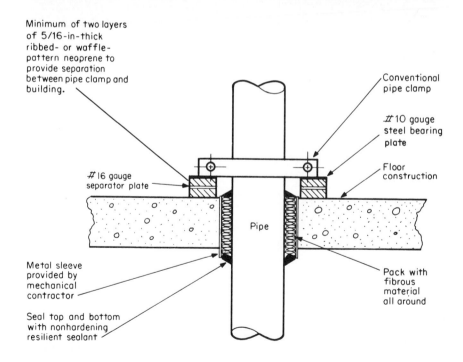

Note:
Neoprene should be 40 to 50 durometer, 40 to 50 psi maximum loading, sized to accommodate the calculated load.

FIG. 4.2 Floor penetration detail for pipes.

   Ribbed or waffle-pattern neoprene pads were recommended for vibration isolators, to be installed according to the detail in Figure 4.2. An alternate method, which usually applies to larger diameter pipes, is shown in Figure 4.3. However, the contractor thought the type of installation used in Figure 4.1 could reduce costs significantly and bolted angle iron clips to the floor and secured threaded rod U bolts through these clips around the pipe. Once the ends of the threaded U-bolt connectors were tightened, the threads on the rod easily cut through the tape, making direct contact with the pipe. As a result, there was no vibration isolation of the pipes.

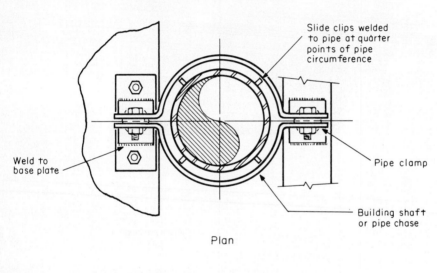

Plan

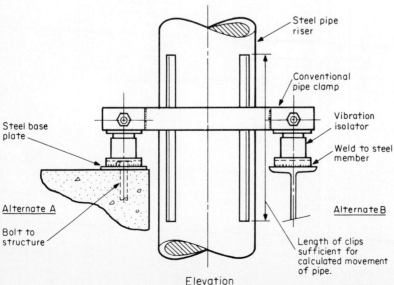

Elevation

FIG. 4.3   **Recommended detail for vibration isolation of anchored vertical pipe riser.**

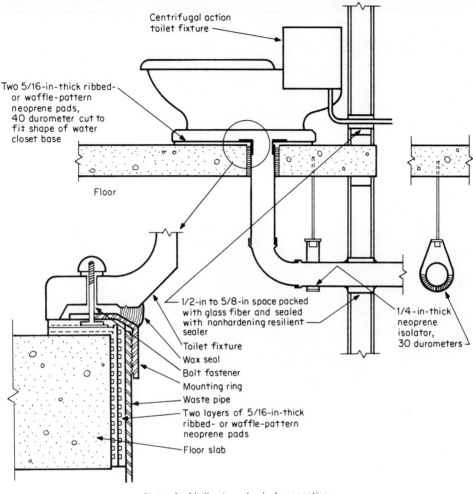

Centrifugal action
toilet fixture

Two 5/16-in-thick ribbed-
or waffle-pattern
neoprene pads,
40 durometer cut to
fit shape of water
closet base

Floor

1/2-in to 5/8-in space packed
with glass fiber and sealed
with nonhardening resilient
sealer

Toilet fixture

Wax seal

Bolt fastener

Mounting ring

Waste pipe

Two layers of 5/16-in-thick
ribbed- or waffle-pattern
neoprene pads

Floor slab

1/4-in-thick
neoprene
isolator,
30 durometers

Note: Avoid direct mechanical connection
between main structure and toilet
fixture and associated plumbing

FIG. 4.4   Vibration isolation for floor-mounted toilet fixture.

To remedy this problem, a compromise was worked
out with the contractor, who was allowed to wrap each
pipe at each floor with two layers of 5/16-in-thick ribbed
or waffle-pattern neoprene pads and then carefully
secure the U bolts so that they were fully separated from
the pipes. This method adequately secures pipes of this
size. The compromise solution (1) provided the required
vibration isolation and (2) allowed the isolators to serve as
diaelectric separators between the two dissimilar metals
of the copper pipe and the threaded steel rods.

Figure 4.4 details the method for vibration-isolating
toilets and the piping connected to them.

4.5 Sanitary Line in Concert Hall Wall

In Figure 4.5, the sanitary or hydraulic line, passing through a noise-sensitive area of a concert hall, was correctly vibration-isolated with the hanger shown. However, when the concrete block wall was later installed, the pipe and the adjacent duct were cemented into the wall, causing the vibration isolators to be ineffective.

All penetrations of the building structure by pipes, ducts, or conduits should receive full and careful acoustical treatment. Such treatments are shown in Figures 3.91 and 3.92.

FIG. 4.5 Sanitary line in concert hall wall.

4.6 Shaft for Ducts and Pipes

Figure 4.6 is a view looking down a masonry shaft for ducts and pipes. At present, only a cast-iron roof leader is installed.

Because the roof leader is very close to noise-sensitive space requiring an acoustical environment of NC-15, a layer of resilient neoprene was recommended for installation between the pipe and the pipe clamps. The neoprene strip, correctly installed, can be seen in the figure. However, the neoprene isolation was ineffective, because the hub of the soil pipe was allowed to touch the wall of the masonry shaft at several places, including the one pictured here. At each point where the pipe touched the wall, the supports for the roof leader had to be adjusted outward, away from the shaft wall, to permit adequate clearance between the soil pipe hubs and the building structure.

FIG. 4.6 **Shaft for ducts and pipes.**

4.7 Vibration-Isolated Soil Lines
4.8 and Water Pipes

Strips of felt padding can be used effectively to isolate certain vibrating pipes and conduits from building structures. Also available for this purpose are manufactured devices such as the one shown in Figure 4.7, properly and neatly installed. The device in this picture is also shown in Figure 4.8.

These units consist of a processed hair felt pad bonded to the inside of a galvanized or cadmium-plated steel outer sleeve. The hair felt pad provides the resilient, sound-absorbing, separating element between the vibrating pipe and the static pipe support.

These devices are generally recommended for soil lines (as shown), small hot-and cold-water plumbing and hydraulic pipes, electrical conduit, and the like in vibrating or noise-carrying pipe systems that pass near or through acoustically sensitive spaces.

FIG. 4.7   Vibration-isolated soil line.

Trisolator — PR –isolator

Function:
1. Sound
2. Vibration
3. Electrolysis

Ribbing

Drop
center

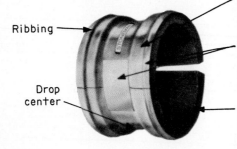

Design and construction features:

Cadmium plated shell: Heavy die formed steel with reinforcing ribs engineered to support the weight of the piping and its contents.

Projecting ribs and drop center: These prevent trisolator from being dislodged from its continuous piano type hinge. Hinge design aligns the halves and prevents deforming the shell during installation.

Isolating pad: Specially processed nonconducting hair felt padding which effectively dampens sound and vibration and reduces their transmission to the building structure. It is of sufficient thickness to provide a wide enough space between the pipe and its support to control electrolytic action. Felt is coated and chemically treated to resist the effects of moisture, abrasion, cold and heat and to repel rodents, insects, etc.
Available in pipe sizes 1/8 through 12 in.

Roller
plates

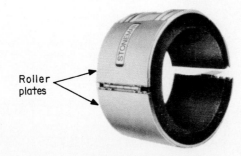

FIG. 4.8 Pipe vibration isolators. *(Stoneman Engineering and Manufacturing Co., top; Potter-Roemer, Inc., bottom.)*

4.9 Pipe Isolation Hangers for Sanitary Line

The hangers shown in Figure 4.9 are cantilevered off the wall, because the horizontal sanitary line was close to the floor. When sanitary lines are located nearer the ceiling, or from the slab above, they may be hung without the cantilever bracket.

FIG. 4.9   Pipe isolation hangers for sanitary line.

4.10 Isolation System for Hydraulic Piping in Concert Hall

In Figure 4.10, the channel-iron pipe support bracket is resiliently isolated from the building structure, which connects to the stage area of a large concert hall in Venezuela. The isolators are manufactured neoprene elements with a durometer of about 40. This system is used for isolating vibrating pipelines from the building structures. These pipes come from hydraulic stage lift pumps.

FIG. 4.10   Isolation system for hydraulic piping to stage lifts in a concert hall.

Figures 4.11 and 4.12 show two different but effective means of providing resilient isolation for plumbing and chilled water distribution lines. The chilled waterlines shown in Figure 4.11 are vibration-isolated from structural steel supports by a neoprene element isolator hanger. The bottom hanger rod is attached to split metal clamps around wood spacer blocks that prevent damage to the thermal insulation. The insulation abuts both sides of the wood spacers when it is installed. Both chilled waterlines are vibration-isolated by one isolator. This is acceptable if care is taken not to overload neoprene elements.

The domestic plumbing or hydraulic line shown in Figure 4.12 is effectively vibration-isolated. The isolating element is a cut section of waffle-pattern neoprene pad, about 5/16 in thick, wrapped around the pipe twice and secured at each pipe clamp. This method is used successfully throughout this large theater and concert hall project in Melbourne, Australia.

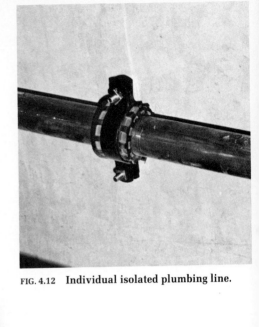

FIG. 4.12   **Individual isolated plumbing line.**

FIG. 4.11   **Plumbing lines racked on wall.**

4.13 Sanitary Piping with Hubless Fitting and
4.14 Resilient Pipe Clamps _____
4.15

Much of the soil and waste piping installed in projects is joined together with "hubless" fittings such as those shown in Figure 4.13.

Interestingly, from a vibration-isolation point of view, these fittings come with a uniquely designed inner sleeve of neoprene material. The neoprene sleeve has an integrally molded inside separator ring against which the joined ends of pipe sections are seated. This ring provides a resilient separation between joined sections.

Where it is desirable to use standard hub-type sections, a neoprene compression gasket may be used at each hub to provide a similar degree of vibration isolation between joined sections. See Figure 4.14 for additional details.

Cushion-clamp assemblies reduce shock and vibration caused by fluid surges in pipes, tubes, and hoses used in the construction of both stationary and mobile equipment. The cushion, made of a special plastic composition, eliminates metal-to-metal contact on machinery piping installations. The special elastomer used in the cushions remains flexible down to −65°F, and is able to withstand temperatures up to +250°F. The cushion material is impervious to most oils, solvents, fuels, and chlorinated fluids.

Smaller plumbing pipes may be installed with a degree of vibration isolation not normally achieved by using

(text continues on page 189)

FIG. 4.13 **Sanitary line using hubless fittings**

Gasket · Spigot without bead · Reinforcing on hub · Hub

Compression joint

Gasket · Stainless steel retaining-clamp · No-hub pipe · Stainless steel shield

No-hub joint

Soundproofing Qualities of Cast Iron with Rubber Gasket Joints

One of the most significant features of both the compression and ¢NO-HUB joints is that they assure a quieter plumbing drainage system. The problem of noise is particularly acute in multiple dwelling units, and although soundproofing has become a major concern in construction design, certain plumbing products have been introduced which not only carry noise but in some cases actually amplify it. The use of rubber gaskets and cast-iron soil pipe reduces noise and vibration to an absolute minimum. Because of the weight and wall thickness of the pipe, sound is muffled rather than transmitted or amplified, and the rubber gaskets separate the lengths of pipe and the units of fittings so that they cushion any contact-related sound. The result is a home that is more livable and of greater value. A detailed discussion of the soundproofing qualities of cast-iron soil pipe DWV systems is contained in Chapter V of Cast-Iron Soil Pipe and Fitting Handbook.

FIG. 4.14   Isolators for hub and No-Hub soil pipe. (*Cast Iron Soil Pipe Institute.*)

Of all the acoustical problems that plague the builder and designer, plumbing noise is among the most serious.

To combat this problem the Institute has developed two significant breakthroughs in "quiet" waste pipe. In one system, a resilient neoprene gasket is used in place of the conventional lead-oakum joint in cast-lion soil pipe. This neoprene gasket prevents any direct metal-contact between lengths of pipe and fittings, and thus establishes a vibration isolation break at every point.

The second system, a somewhat more sophisticated one, is known as the hubless system. Develop by the Cast Iron Soil Pipe Institute, this system uses hubless (plain end) pipe joined together in a neoprene sleeve which is secured in an outer band of stainless steel.

1. Hubless cast iron soil pipe ready for joining. Note the extreme simplicity of joint components.

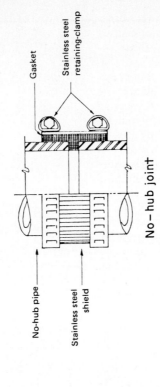

2. Sleeve gasket is placed on end of one pipe length. Stainless steel shield with attached band clamps is placed on end of other pipe.

3. Pipe ends are butted squarely against integrally molded cushion (separator ring) inside of gasket. Joint is quickly assembled and permanently fastened.

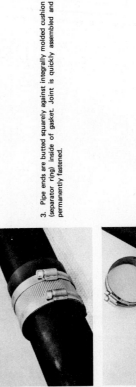

4. Fittings are jointed in the same way—simplified and easily done in close quarters. The protection shield is rotated for most convenient access for torquing the clamp screws.

cushion clamps similar to those shown on Figure 4.15.

Installing the Hydra-Zorb clamping system is fast and easy. Standard steel or aluminum channel, either single or double, is used as the base for the system.

Once the channel is in place, one person using one simple tool can complete the job. Only one cushion and clamp are required for each tube size, and predetermined spacing is not critical. In addition, Hydra-Zorb cushion-clamp assemblies allow fluid conductors to be added to or removed from installations without disturbing adjacent tubing or piping.

Hydra-Zorb cushion-clamp assemblies are available for steel tubing sizes from ¼- to 3⅛-in, o.d. and for steel piping sizes from ¼- to 6-in i.d. Cushion-clamp assemblies for hydraulic hose are obtained by measuring the hose outer diameter and selecting the closest standard size.

For additional information and technical data for soil pipe systems refer to *Noise and Vibration Characteristic of Soil Pipe Systems*, Polysonics, Washington, D.C., and *Cast Iron Soil Pipe & Fittings Handbook*.

Hydra-Zorb cushion-clamp assemblies reduce shock and vibration caused by fluid surges in pipes, tubes, and hoses used in the construction of both stationary and mobile equipment.

The Hydra-Zorb cushion, made of a special plastic composition, eliminates metal-to-metal contact on machinery piping installations. The special elastomer used in the cushions remains flexible down to −65° F. and is able to withstand temperatures up to +250° F. The cushion material is impervious to most oils, solvents, fuels, and chlorinated fluids.

FIG. 4.15 **Isolators for small-diameter pipes.** *(Hydra-Zorb.)*

PLUMBING PUMPS

4.16 Vibration Isolation for Plumbing Pumps

Depending on the type of building and the importance of controlling noise and structure-borne vibrations, it may be necessary to provide vibration isolation for certain pieces of plumbing equipment. Such vibrations may be heard as radiated noise at locations in the building structure remote from noisy equipment.

In concert halls or music buildings, where a submersible pump is required, a vibration-isolation system similar to that shown in Figure 4.16 may be used successfully.

In this detailed arrangement for controlling noise and vibration, not only have the pump and direct connected sections of piping been isolated from the building, but a specially designed flexible piping connector has been used. The flexible connector here must be more than a device to prevent misalignment, such as a section of moulded and reinforced rubber, a metal bellows, or straight neoprene sleeve. The connector should be relatively lightweight and be able to expand spherically and function without the use of control rods. Additionally, the unit should be able to operate more like an automobile tire so that the connector "sees" fluid pulsations. Like an automobile tire, the unit will be able to expand and contract from the effects of the fluid pulsations, which tend to dampen these pressure variations, thus providing a degree of attenuation of fluidborne vibration not previously attainable in more conventional units.

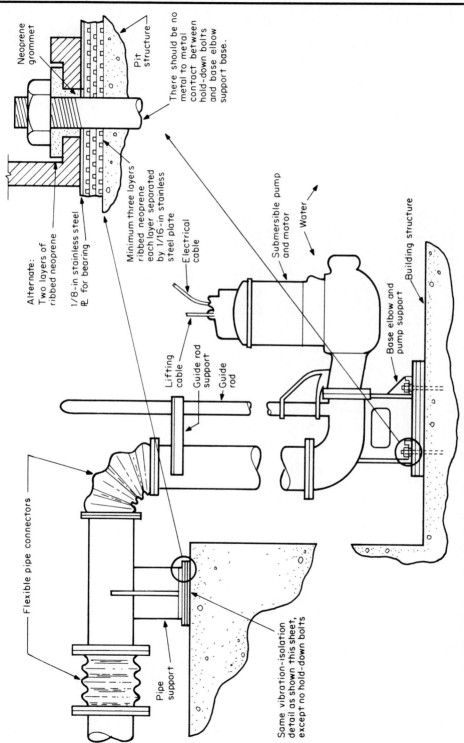

Neoprene grommet

Pit structure

There should be no metal to metal contact between hold-down bolts and base elbow support base.

Alternate:
Two layers of ribbed neoprene

1/8-in stainless steel PL for bearing

Minimum three layers ribbed neoprene each layer separated by 1/16-in stainless steel plate

Electrical cable

Submersible pump and motor

Water

Building structure

Lifting cable

Guide rod support

Guide rod

Base elbow and pump support

Flexible pipe connectors

Pipe support

Same vibration-isolation detail as shown this sheet, except no hold-down bolts

FIG. 4.16   Recommendations for vibration isolation of submersible pump.

4.17 Pipeline Vibration Isolators

These flexible pipe connectors may be installed in either vertical or horizontal positions. The manufacturer's preferred location when only one connector in each pipe is used is to have it parallel to the equipment shaft as equipment vibration tends to be more severe in a direction radiated to the shaft.

The maximum isolation connectors may be used both vertically and horizontally in the same pipeline as seen in Figure 4.17.

Elbow and straight-through flexible fixtures are available for solving vibrating directional problems where piping must make a right angle turn as it leaves and enters equipment.

FIG. 4.17   Pipeline vibration isolators.

4.19 This installation is representative for plumbing (hydraulic) pumps such as sump, sewage, and ejector pumps. Neoprene grommets like those shown in Figure 3.26b and 3.26c, along with neoprene washers, were used to secure the steel base plate of the pump to the concrete floor. Pump discharge pipes were resiliently isolated, as shown in Figure 4.9. The entire duplex pump unit was eventually vibration-isolated with two layers of neoprene strips placed under all four sides of the base plate. (This treatment is not shown in Figure 4.18.)

An acceptable alternative to this treatment would be to provide two layers of neoprene cut in circular pads, to fit under the mounting frame of each pump, between the pump's base and the square steel base plate (see arrow.)

Figure 4.19 shows a recommended method for vibration-isolating this type of equipment and its associated piping.

FIG. 4.18 Sewage ejector and sump pump.

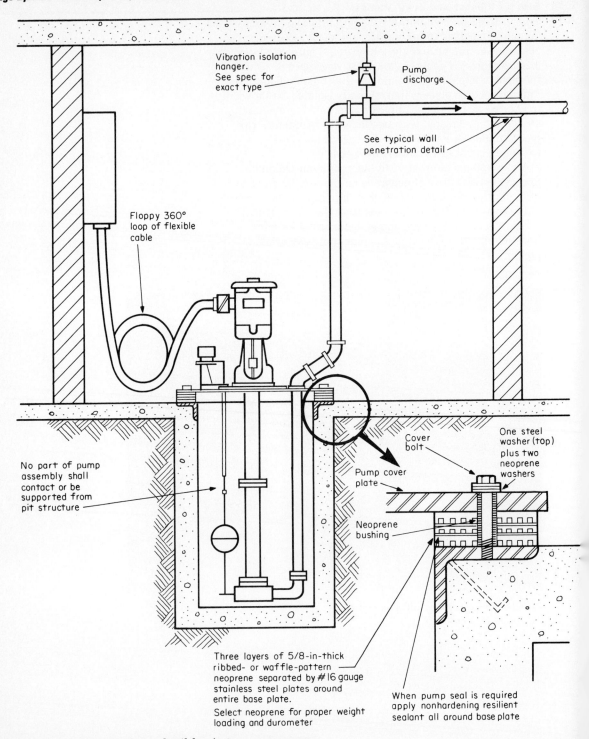

Vibration isolation hanger. See spec for exact type

Pump discharge

See typical wall penetration detail

Floppy 360° loop of flexible cable

No part of pump assembly shall contact or be supported from pit structure

Cover bolt

One steel washer (top) plus two neoprene washers

Pump cover plate

Neoprene bushing

Three layers of 5/8-in-thick ribbed- or waffle-pattern neoprene separated by #16 gauge stainless steel plates around entire base plate.

Select neoprene for proper weight loading and durometer

When pump seal is required apply nonhardening resilient sealant all around base plate

FIG. 4.19 Vibration-isolation detail for ejector or sump pump.

4.20 In-Line Pumps

On the two in-line pump circulators in the foreground of
Figure 4.20, the supply and return piping has been
adequately isolated by vibration isolators (not visible in
the figure). The full 360° loops of armored conduit have
been correctly installed to provide the necessary resilient
connection between pump motors and the rigid conduit
sections attached to the building.

However, the hard-mounted, rigid steel pipe stanchions
located under each pump have not been installed
correctly for effective noise and vibration control. Because they are attached directly to the floor, vibrations
from the pump are transmitted into the floor slab.

This problem was solved by installing two layers of
ribbed neoprene under the base plate (see arrow) of each
stanchion, thereby providing the required vibration
isolation.

FIG. 4.20 In-line pumps.

Figure 4.21 shows an actual installation of the recommended vibration-isolation detail depicted in Figure 4.19 for sump or ejector pumps. Note the several layers of ribbed neoprene pad strips, which extend around the entire four sides of the pit cover/pump support plate.

For vibration isolation of the piping systems, flexible pipe fittings may be used at piping connections to the pumps and the pit cover plate. An acceptable alternative to the above is to provide vibration-isolation hangers for the connected piping systems. Generally, this arrangement should include each hanger support for a distance of at least 50 ft from the pumps. In more noise-sensitive or acoustically critical applications, each project should be dealt with separately and in such detail as to ensure satisfactory results.

FIG. 4.21 **Sump or ejector pumps.**

4.22 Sump-Pump Isolation

Figure 4.22 shows one method of vibration-isolating a sump pump. Two layers of neoprene pads, sized to accept the proper weight load, are placed under each hold-down bolt. Bolts are fitted with a neoprene sleeve where they pass through the pump base. Neoprene washers are installed beneath each steel washer under each hold-down nut. This treatment provides for a completely resilient securement.

The spaces between the pump base and the floor and between the isolators were resiliently sealed with strips of soft rubber and sealed with a bead of nonhardening, resilient, waterproof caulking.

FIG. 4.22   Sump-pump isolation.

Electrical Conduit and Conductors

5

5.1 Electrical Conduits

All penetrations of walls, floors, or slabs of construction housing noisy equipment should be treated acoustically to prevent noise from being transmitted out of the noisy space. In the case of electrical conduits, and where vibration transmission is not a major concern, the conduits may be grouted or cemented in place in the penetration, as shown in Figure 5.1. The slightest movement between poured concrete elements and concrete block walls will produce slight cracks. The normal shrinkage of cement grout will also produce small cracks. To maintain a permanent seal where these cracks occur, a bead of nonhardening resilient sealer should be used. The same sealing should be used around conduits.

FIG. 5.1   Electrical conduits.

5.2 Electrical Conductor Penetration

5.3 Manufactured devices are available for multiple electric
5.4 cable penetrations through building construction.
5.5 Descriptive information and details from two manufacturers are shown in Figures 5.2 to 5.5.

Figure 5.2 shows the easy three-step installation for
electrical conductor penetration:

1. Figure 5.2*b*. When pouring concrete during wall or
floor construction, cast in the one-piece mounting frame
or grout frame into existing masonry surfaces, whichever
is most convenient for the particular installation.

2. Figure 5.2*c*. Feed cables through the frame. Insert
factory-assembled sealing block into mounting frame's
keyway and slide forward until stops on assembly meet
edge of frame. Assembly is automatically aligned
vertically and horizontally.

3. Figure 5.2*d*. Position each cable in a sealing block
opening of corresponding size. Using standard box
wrench or deep socket with ratchet or drill, tighten bolts
on clamping hardware.

Figure 5.3 shows another system of electrical conductor
penetration. *Mounting frames* are available in sizes to
accommodate a wide range of cable tray sizes and
loadings, including single and multiple layers of cables
for power or instrument applications. Cast keyways in
mounting frame align and position sealing block assemblies for precise fit. Frames can be installed in wall such
that sealing block assemblies can be inserted in either
horizontal or vertical position.

(b)

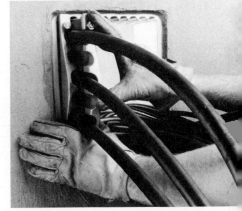

(c)

(a)

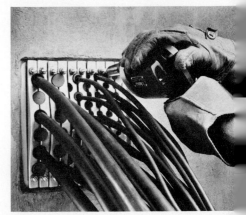

(d)

FIG. 5.2   Electrical conductor penetration closure. *(Crouse-Hinds.)*

The *sealing block assembly* consists of a specially formulated elastomeric material between cast-malleable-iron pressure plates. It protects cable from mechanical damage and provides high pullout resistance while also ensuring positive cable separation. The elastomeric material expands during fire to seal any voids left by burned cable insulation. Cast stops on the front pressure plate prevent the sealing block assembly from slipping through the mounting frame during installation. Assemblies are offered for all cable/conduit outside diameters from 0.250 to 2.250 in (6.4 to 57.2 mm). Sealing block openings will accommodate undersize and out-of-round cables. Each sealing block assembly seats multiple

(a)

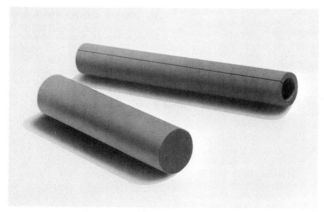

(b)

FIG. 5.3 Electrical conductor penetration closure. *(Crouse-Hinds.)*

Multi-Cable
Transit

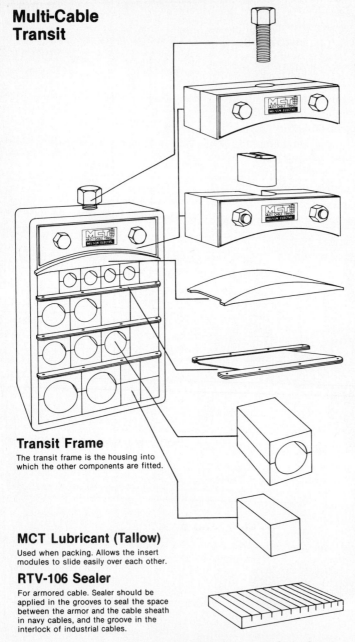

Compression Bolt

When tightened, the bolt applies pressure to the compression plate sealing the grooved insert modules around the cables.

End Packing — Standard

End packing assembly is bolted into place to provide a fire and watertight seal above the compression plate. The standard end packing assembly is used when both sides of the transit frame are accessible.

End Packing — Special

The special end packing assembly serves the same purpose as the standard and is used when the transit frame is accessible from only one side.

Compression Plate

The compression plate acts as a pressure plate above the internal assembly.

Stay Plates

Stay plates are inserted between every completed row to help distribute compression forces within the frame and to keep modules from dislodging under high pressure conditions.

Grooved Insert Modules

Grooved insert modules are available in seven module sizes to accommodate a range of cable/pipe from 5/32″ to 3-3/4″ O.D. They fit snugly around the cable or pipe to form an air-tight, water-tight seal when compression is applied in final assembly step.

Spare Insert Modules

Solid modules are used to fill voids or allow for future addition of cables. They are available in 3 module sizes.

Fill-In Insert Strips

Used to fill space gaps. Available in two thicknesses: 5 and 10 mm. Strips are 120 mm long and are split to allow cutting at any desired length.

Transit Frame

The transit frame is the housing into which the other components are fitted.

MCT Lubricant (Tallow)

Used when packing. Allows the insert modules to slide easily over each other.

RTV-106 Sealer

For armored cable. Sealer should be applied in the grooves to seal the space between the armor and the cable sheath in navy cables, and the groove in the interlock of industrial cables.

FIG. 5.4   Electrical conductor penetration closure. *(Nelson Electric.)*

cables/conduits. Compact design permits close nesting of cables, saving space.

Reducers permit sealing block assemblies to accept cables with smaller outside diameter than the specified range. *Plugs* fill unused openings in sealing block assemblies. Blank sealing block assemblies (not shown in Figure 5.3) fill unused spaces in the mounting frame, providing for future system expansion.

Neither manufacturer makes any claims about the acoustical ratings for these devices. However, when completely installed and sealed in accordance with the manufacturer's recommendations, the devices do provide an airtight closure and fire barrier. If there are no air paths through the device, there can be no airborne noise transmission through the device. This does not mean that it stops *all* noise; it blocks only that noise that can be carried via an actual air path. Thus, the amount of noise these devices can stop is limited by their mass, like any device used to seal penetrations.

Typical Frame Installation Methods

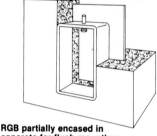

RGB partially encased in concrete for flush mounting

(UL) classified 3 hour wall and floor fire rating

RGS welded on steel plate wall

ANI (NEL-PIA) approved (American Nuclear Insurers) 5 hr. wall and floor rating

RGM surface mounted with bolts and Tecron® gasket

Installation of Multi-Cable Transit is quick, easy, and economical. The basic steps are described below:

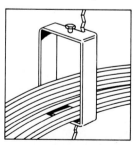

1. Empty frames are cast into or surface mounted to walls or floors by conventional construction methods. Cables, conduit, or pipe are run according to standard design criteria.

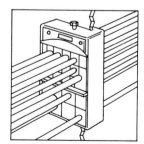

6. Insertion and tightening of end packing completes the job.

FIG. 5.5   Electrical conductor penetration closure. *(Nelson Electric.)*

5.6 Vibration-Isolating Electrical Conduits

Illustrated in Figure 5.6 is an acceptable, effective method for vibration-isolating electrical conduits from the building construction and the wall-mounted electrical distribution panel. The bracket to which the conduit clamps are secured is isolated from the concrete block masonry wall by neoprene pads. These pads are out of sight in Figure 5.6, but clearly identifiable are the black neoprene grommets that provide resilient separation of the conduit where it passes through the top of the panel.

FIG. 5.6   Vibration-isolating electrical conduits.

5.7 Small-Diameter Flexible Conduit

Shown in Figure 5.7 is an effectively installed, small (1-in diameter), full 360° loop of flexible electrical conduit. In small sizes (1 in and less), when looped as shown, these sections of flexible conduit serve as effective vibration isolators. Notice that the loop is self-supporting and does not touch any nearby utility systems or the building structure. The conduit manufacturer's minimum bend radius for each size of conduit should never be exceeded if the conduit is expected to remain flexible and resilient.

The looping of flexible conduit is recommended only for conduits having a diameter of 1 in and less. For vibration-isolating larger diameter conduits, a fitting similar to that shown later in Figures 5.9 and 5.10 may be used.

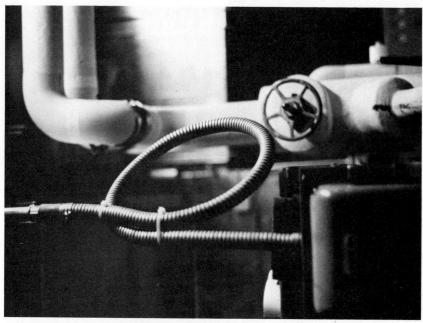

FIG. 5.7 Small-diameter flexible conduit.

5.8 Large-Diameter "Flexible" Conduit

In Figure 5.8, a section of supposedly "flexible" conduit (see arrow), which connects the rigid conduit to the motor box of the electric motor in a cooling tower, is not truly flexible. This particular connection is fabricated from armored flexible conduit with a vinyl weatherproof jacket for out-of-doors applications. The diameter of the conduit is 4 in, and the length of the "flexible" section is about 24 to 30 in, which is far too short to be really flexible. Actual field installations reveal that flexible corrugated metal conduit with a diameter larger than 1 in, even when coiled in a 360° loop not exceeding the manufacturer's minimum bending radius, is not really flexible enough to act effectively as a vibration-isolation device. In reality, the metal corrugations bind on one another, thus providing a straight continuous structural path by which vibrational energy may travel.

Another disadvantage to the use of coiled 360° loops of flexible conduit is that the larger the diameter of the conduit, the larger the loop, to the point where the full recommended bending radius would result in a coil so large that it would not fit into the building space. For example, one conduit manufacturer recommends the parameters shown at the right for flexible conduit, stopping at 5-in diameter.

Conduit diameter, in	Weight, lb/1000 ft	Minimum bending radius, ft
5/16	150	1¾
3/8	250	2
		3
3/4	570	4
	880	5
1¼	1100	6¼
1½	1320	7½
2	1720	10
2½	2600	12½
3	3000	15
3½	3600	17½
4	4000	20
5	—	—

FIG. 5.8 Large-diameter "flexible" conduit.

Preliminary tests indicate that this connector is effective
in most critical frequencies for conventional HVAC
mechanical and electrical equipment installations. In the
large conduit sizes (over 1-in diameter), it is far more
effective than selections or loops of so-called flexible
electrical conduit. Therefore, the coupling shown in Figures 5.9 and 5.10 is recommended for all conduit of
1½-in diameter or larger.

FIG. 5.9 Vibration test setup showing flexible electrical conduit connector.

FIG. 5.10 Close-up view of flexible coupling showing some of the construction details.

5.11 Electrical Cable Racks ⎯⎯⎯⎯⎯⎯⎯⎯⎯⎯⎯⎯⎯⎯⎯⎯⎯⎯⎯⎯⎯⎯⎯⎯

5.12 The electrical cables on the rack must penetrate the poured concrete wall, as shown in Figure 5.11. The wall forms part of the building construction that separates a noisy mechanical equipment room from the very quiet acoustical environment of the concert hall serving one of Australia's major cities.

Unless the open space around the rack and cable penetration of this wall is acoustically sealed, enough mechanical equipment noise can pass through the opening to cause measured sound pressure levels at center stage to reach almost NC-25. That exceeds the normal concert hall design criteria of NC-15 by 10 NC points, or about 10 dB. Normally, the noisy excess would occur at the blade passage frequency of the fan systems; for centrifugal fans, that is usually at 125 Hz. Other types of equipment, such as chillers, compressors, or pumps, could also contribute to excess noise at other frequencies.

FIG. 5.11 **Electrical conductors penetrating a wall.**

It cannot be overemphasized that correct packing and sealing of such penetrations as this are absolutely vital when trying to achieve certain acoustical environments such as required in this concert hall.

A similar situation is that shown in Figure 5.12. Again, the wall being penetrated by cable trays and conductors, and other items, is separating mechanical equipment from a concert hall, this time in Baltimore, Maryland.

One can thus see that problems of this nature occur worldwide, and the laws of physics operate the same with regard to noise transmission whether one is "on top of the world" or "down under."

FIG. 5.12 **Conductors and cable trays penetrating a wall.**

5.14 Closely racked multiple conduits that penetrate building
walls, floors, or ceilings present a rather difficult situation to treat acoustically. The two layers of multiple
electrical conduits in Figure 5.13 clearly show such an
instance.

Each black steel pipe sleeve has been wrapped before
installation with a section of resilient foam rubber.
Figure 5.14 shows specific details.

In arranging this battery of conduits, the contractor has
wisely arranged the conduits in two layers, allowing
room above and below the conduit penetration. Once the
block wall (just visible in the lower portion of Figure
5.13) is closed in, the contractor can easily pack the
space between each conduit and its sleeve with fibrous
material and seal each end of the sleeve with a nonhardening resilient sealer, effectively completing the
acoustical treatment.

FIG. 5.13   **Multiple electrical conduits.**

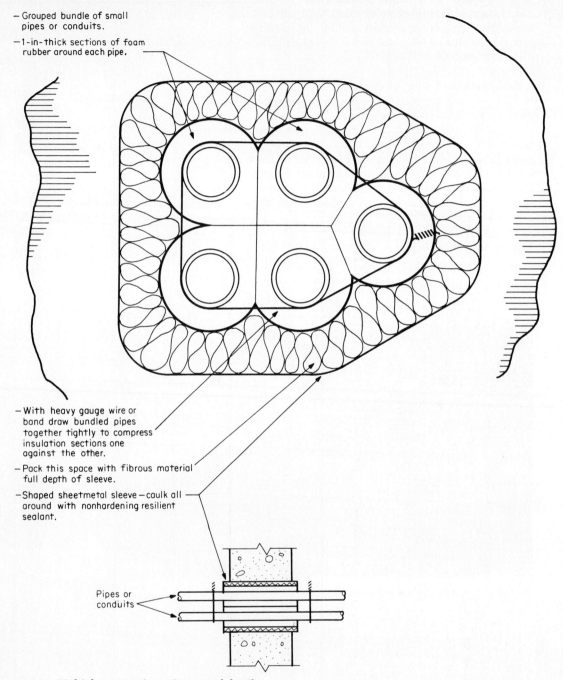

— Grouped bundle of small pipes or conduits.

—1-in-thick sections of foam rubber around each pipe.

— With heavy gauge wire or band draw bundled pipes together tightly to compress insulation sections one against the other.

— Pack this space with fibrous material full depth of sleeve.

—Shaped sheetmetal sleeve—caulk all around with nonhardening resilient sealant.

Pipes or conduits

FIG. 5.14   Multiple-penetration noise control detail.

ELECTRICAL TRANSFORMERS

5.15 Small Transformer — Hung

In most installations, basic inexpensive vibration-isolation systems should be provided for small electric transformers located throughout a project.

Figure 5.15 shows a transformer neatly hung on simple neoprene hanger isolators. If properly selected, these work well to reduce vibration transmission to the building structure. Note the long, slack section of flexible conduit between the transformer and the rigid conduit secured to the building wall. For smaller sizes of conduit, this arrangement is adequate. Larger conduits require a vibration-isolation fitting similar to that shown in Figures 5.9 and 5.10.

The whole vibration-isolation system in Figure 5.15 is far enough away from the wall and nearby systems and the isolation hangers are far enough from the transformer housing that there is no danger of reducing the effectiveness of the isolation system.

FIG. 5.15   Small transformer — hung.

5.16 **Vibration Isolation for Transformer**

5.17
5.18
5.19
There are typical methods for vibration-isolating small, floor-mounted electrical transformers (Figures 5.16 and 5.17). Notice the section of small-size flexible armored conduit and the neoprene waffle pads under the corners of the transformer feet. Thus isolated, the transformers stand clear of the building structure and all adjacent equipment. Neoprene pads should be 40 to 50 durometer. Neoprene mounts similar to those shown in Figure 5.19 are preferred as being most efficient.

Figure 5.18 is used to illustrate what can happen when neoprene pads are subjected to a point load, which is more than the pads can handle. This should show why a steel plate, which is the full size of the pads, should be positioned on top of the pads to correctly distribute the weight load imposed at each isolator.

FIG. 5.18   **Overloaded neoprene pad isolators.**

FIG. 5.16   **Small transformer — floor mounted.**

FIG. 5.17   **Floor-mounted transformer.**

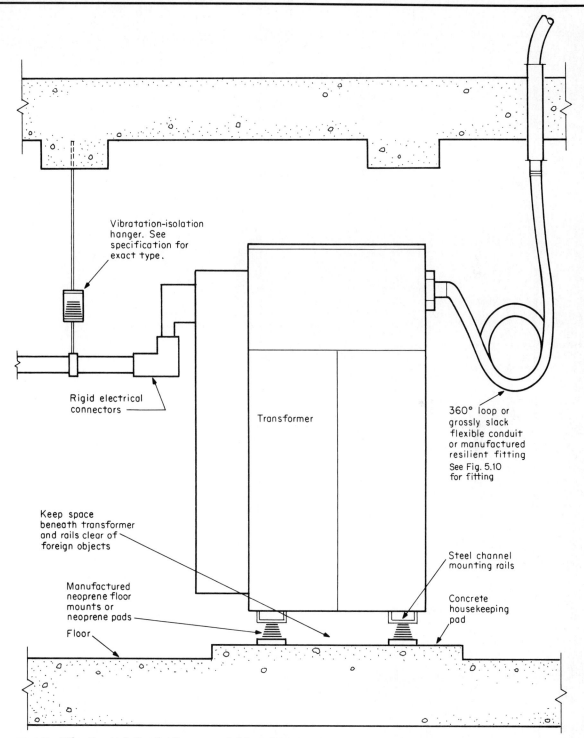

Vibratation-isolation hanger. See specification for exact type.

Rigid electrical connectors

Transformer

360° loop or grossly slack flexible conduit or manufactured resilient fitting See Fig. 5.10 for fitting

Keep space beneath transformer and rails clear of foreign objects

Steel channel mounting rails

Manufactured neoprene floor mounts or neoprene pads

Concrete housekeeping pad

Floor

FIG. 5.19   Vibration isolation for floor-mounted transformers.

5.20 Transformer Duct Connection

In the installation shown in Figure 5.20, a battery of
electrical conductors was installed inside a fabricated
sheet-metal duct. The conductors run up from the
transformer shown at the bottom of the figure to a standard electrical distribution box (not shown). To make the
connection of the duct to the transformer resilient, a
neoprene material was used, forming a skirt (see arrow)
around all four sides of the duct. Duct support hangers
were provided with neoprene pads. (See Figure 5.21.)

FIG. 5.20   Transformer duct-bank connections.

5.21 Transformer Duct Isolators

The arrow in Figure 5.21 points to the vibration-isolation pads at the channel-iron support members for the electrical conductor duct. Neoprene pads may be ribbed or waffle-pattern in two layers, each approximately $5/16$-in thick. Generally, the durometer for the neoprene pads should be from 40 to 50.

In such installations, the conductor duct should be kept completely free from the building structure and all nearby utility systems. A neoprene skirt similar to the one shown in Figure 5.20 should also be provided at the connection where the conductor duct joins the sheet-metal distribution box.

FIG. 5.21   Transformer duct-bank isolators.

5.22 Portable Transformer Isolation

Figure 5.22 shows an installation of caster-mounted mobile electrical transformers for a project in Australia.

The transformer casters are placed in pipe flanges of selected sizes to fix the units in the position and to help distribute the transformer weight. In addition, steel plates, properly sized to further distribute the weight evenly over the ribbed neoprene pad isolators, are clearly evident in the figure. This rather simple vibration-isolation system is very effective in satisfactorily controlling the high-frequency vibrations emanating from the transformer units.

FIG. 5.22 **Portable transformer isolation.**

5.23 Fixed Transformer Isolation

Figure 5.23 shows an acceptable method of isolating a permanently fixed transformer. It also shows an acceptable method of grounding; however, it should be noted that the neoprene element in the vibration isolator is badly overloaded.

FIG. 5.23 **Fixed transformer isolation.**

5.24 Flexible Connections to Transformers

5.25 Connections to vibrating electric transformers generally need to be flexible and resilient. When sections of loosely fitting flexible conduit are used, as in Figure 5.24, it is not recommended that the piece of flexible, vibration-isolating conduit be attached midway to the wall or other section of the building. The entire section of flexible conduit should be left unsecured for its entire length.

On the other side of this same transformer, sections of fabricated sheet metal were used to form a duct through which the electrical conductors could pass to the metal distribution panel (far right of Figure 5.25), which was secured firmly and without resilient means to the building wall. Such a ducting arrangement, unless resiliently separated from the transformer, provides a bridge or path by which vibration energy passes onto and on through the building construction. In this case, the bottom section of duct, fastened to the transformer, and the top section, screwed to the distribution box, were finally separated in between with a foam rubber pad, which can be seen in the figure (see arrow). The horizontal section of duct extending out from the distribution box had to be shortened so that it could not directly touch the transformer housing.

FIG. 5.24 Flexible conduit—secured.

FIG. 5.25 Sheet-metal duct connectors from transformer to distribution panel.

EMERGENCY ELECTRIC GENERATORS

5.26 Emergency Electric Generator

Figure 5.26 shows that emergency electric generators may be safely and satisfactorily point-isolated on open, stable, freestanding spring isolators. Note that the generator exhaust piping is isolated on spring hangers. This degree of vibration isolation for a piece of equipment that will operate only at times of emergency and strategically scheduled exercise periods may or may not be necessary, depending on its location, the type of building, and the owner's requirements.

FIG. 5.26   Emergency electric generator.

5.27 Emergency Electric Generator

5.28
5.29 The emergency electric generator in Figure 5.27 is mounted on housed isolators. When the engine is not operating, the isolation units generally stay upright and in alignment. Once the engine is started, however, the units often go out of alignment and malfunction. The weight of the machine and the unbalanced operating forces of the engine cause the components of the isolators to pound against each other, transmitting considerable vibrational energy to the building structure.

Resilient snubbers should have been installed to limit or restrict the equipment movement during operation, and open, unhoused isolators should have been used in place of those shown.

Similar applications of generator vibration isolation can be seen in Figures 5.28 and 5.29 in other locations in the United States and Canada.

FIG. 5.27   Emergency electric generator.

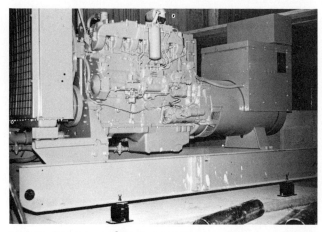

FIG. 5.28   Emergency electric generator.

FIG. 5.29   Emergency electric generator.

DIMMERS

5.30 Recommended Vibration Isolation for
5.31 Dimmer Bank _____

5.32 The vibration isolation usually required and recom-
5.33 mended for a dimmer bank is detailed in Figure 5.33,
and Figures 5.30 and 5.32 show such applications.

The ribbed or waffle-pattern neoprene pads, usually
two layers of 5/16- to 3/8-in thickness each, are sized and
positioned at locations under the dimmer base to accept
the proper load so as to function as designed. The
neoprene durometer should usually be 30 to 40.

Connections to the dimmer bank should be made with
sections of loosely coiled flexible conduit or special
vibration-reducing couplings similar to that shown in
Figures 5.9 and 5.10.

If the dimmer bank is quite tall and engineers feel that
some means of securement is needed near the top of the
unit, such connections should be made resiliently with
the same type of neoprene pads that are used under the
dimmer base. See Figures 5.31 and 5.33 for recommended
installation and detail.

FIG. 5.30   Dimmer bank isolator.

FIG. 5.31   **Resilient connections to dimmer bank.**

FIG. 5.32   **Isolation pads under dimmer.**

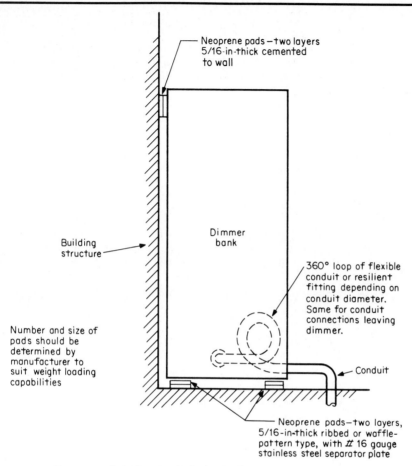

Neoprene pads—two layers
5/16-in-thick cemented
to wall

Building
structure

Dimmer
bank

360° loop of flexible
conduit or resilient
fitting depending on
conduit diameter.
Same for conduit
connections leaving
dimmer.

Number and size of
pads should be
determined by
manufacturer to
suit weight loading
capabilities

Conduit

Neoprene pads—two layers,
5/16-in-thick ribbed or waffle-
pattern type, with #16 gauge
stainless steel separator plate

FIG. 5.33 **Recommended vibration isolation by dimmers.**

Walls, Floors, Ceilings, and Doors

6.1 Conventional Methods for Controlling the Transmission of Airborne Excitations

Noise and vibration travel from noisy spaces to acoustically sensitive spaces through walls, ceilings, and floors. Figure 6.1 shows the noise paths that are most common in building construction. Possible noise control treatments are shown for each of these paths. Further discussion of these treatments is provided in each of the categories of this section.

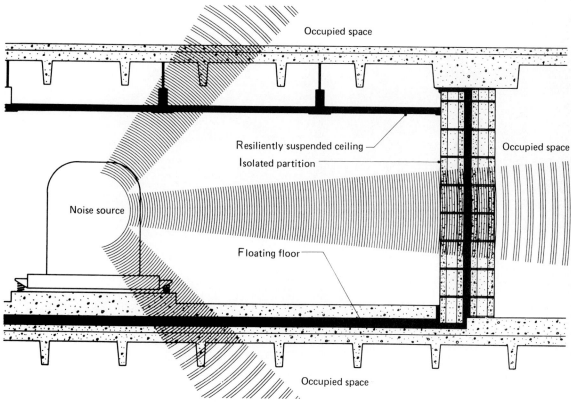

Occupied space

Resiliently suspended ceiling

Isolated partition

Occupied space

Noise source

Floating floor

Occupied space

FIG. 6.1   Noise source and noise paths. *(Peabody Noise Control.)*

WALLS

6.2 Resiliently Mounted Wall System _____

6.3 Walls in buildings are usually resiliently isolated to improve the sound transmission class (STC) between adjacent spaces or to prevent flanking of noise around floated floors. The term "flanking" is used to describe a noise path that goes around an isolated component.

An example of a resiliently isolated wall used in conjunction with a floated floor is shown in Figure 6.3. Figure 6.2 shows an actual installation under construction for a rehearsal room at a large concert hall in Caracas, Venezuela.

FIG. 6.2   Resiliently mounted wall system.

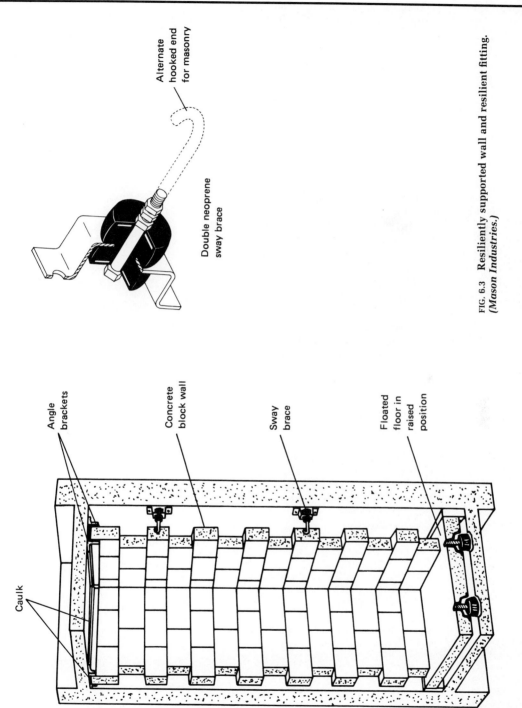

Alternate hooked end for masonry

Double neoprene sway brace

FIG. 6.3 Resiliently supported wall and resilient fitting. (*Mason Industries.*)

Angle brackets

Concrete block wall

Sway brace

Floated floor in raised position

Caulk

6.4 Wall Treatment in Mechanical Equipment Room

Figure 6.4 shows typical construction for the noise control treatment in a mechanical equipment room housing an air-handling unit. The room is above and to one side of a noise-sensitive meeting room. A wall designed and constructed so as to achieve a background sound of NC-30 to NC-35 in the meeting room separates it from the mechanical equipment room, in this stadium project in New Orleans. Metal studs erected in a staggered pattern separate the inner wall of the mechanical equipment room from the outer wall, next to the meeting room ceiling space. To provide some improvement in isolation, 2 in of glass fiber insulation was applied in the wall cavity between the studs. The inner and outer skins of the wall were constructed of two layers of ⅝-in-thick gypsum board. All joints were staggered and sealed airtight.

A key element in the success of this installation is shown in Figure 6.4 where the wall meets the irregularly shaped corrugated steel deck. The contractor carefully cut, fit, and sealed the individual pieces of gypsum board around the steel deck corrugations. To make the entire application airtight and noiseproof, a bead of sealing compound was later applied at the abutting joint of each of the small pieces of gypsum board and all along the bottom horizontal joint. The completed wall performed acoustically as predicted.

The duct penetrating the wall was also packed and sealed acoustically and mechanically vibration-isolated to prevent ductborne vibrations from turning the wall into a noise-reradiating panel.

FIG. 6.4 **Wall treatment in a mechanical equipment room.**

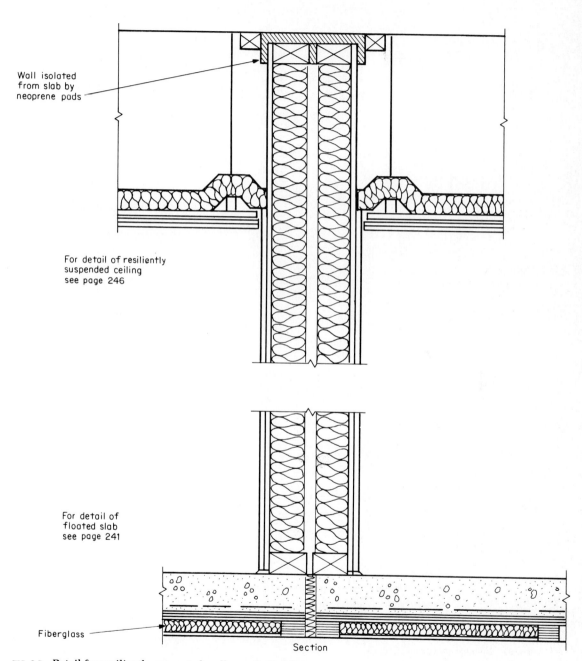

Wall isolated
from slab by
neoprene pads

For detail of resiliently
suspended ceiling
see page 246

For detail of
floated slab
see page 241

Fiberglass

Section

FIG. 6.5   Detail for resiliently supported wall on a floated floor.

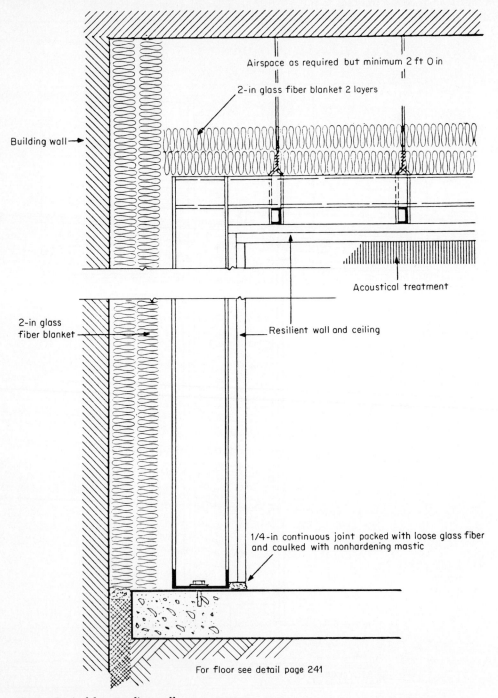

Airspace as required but minimum 2 ft 0 in

2-in glass fiber blanket 2 layers

Building wall →

Acoustical treatment

2-in glass
fiber blanket →

Resilient wall and ceiling

1/4-in continuous joint packed with loose glass fiber
and caulked with nonhardening mastic

For floor see detail page 241

FIG. 6.6 **Typical freestanding wall.**

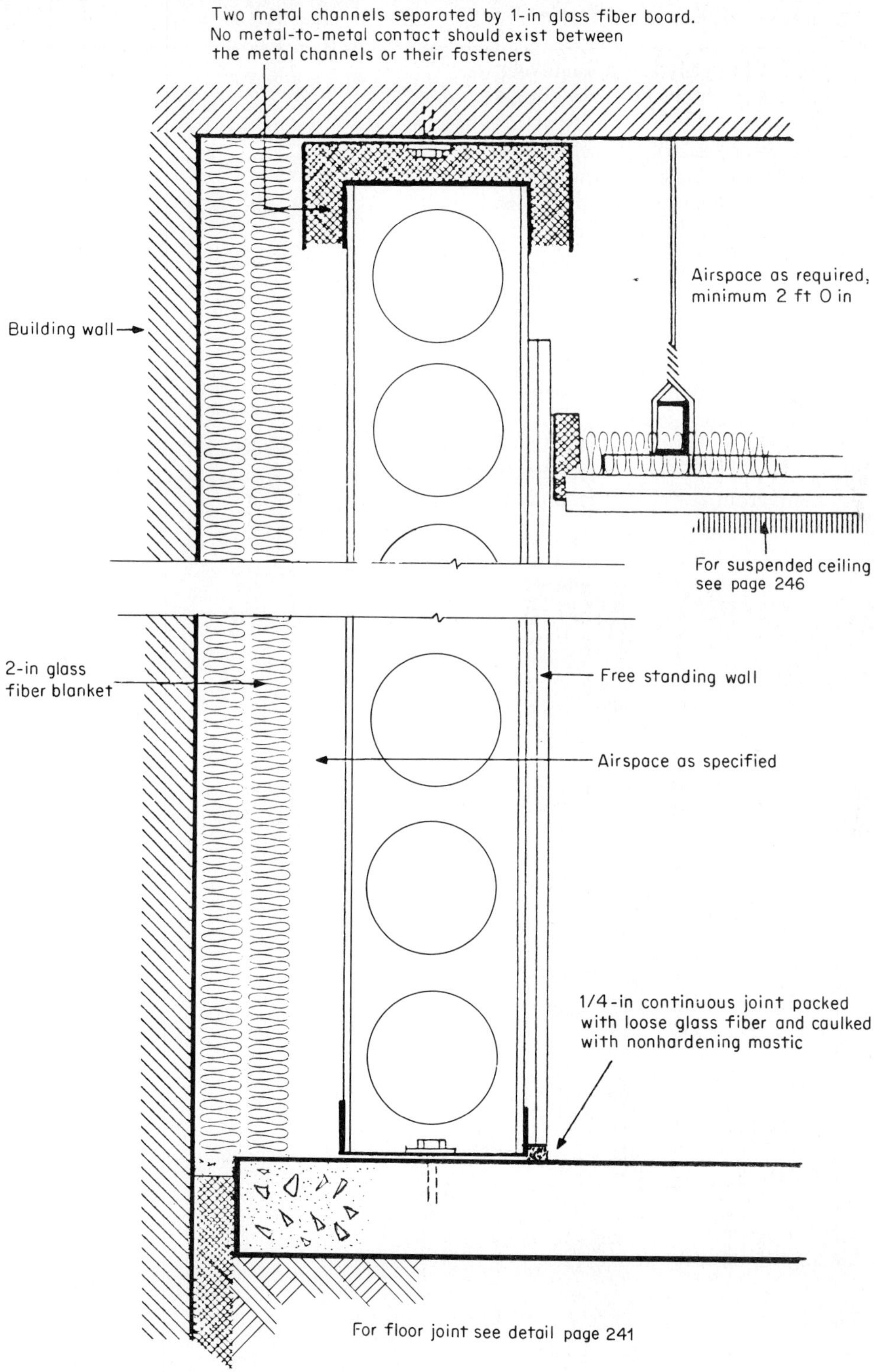

Two metal channels separated by 1-in glass fiber board. No metal-to-metal contact should exist between the metal channels or their fasteners

Building wall →

Airspace as required, minimum 2 ft 0 in

For suspended ceiling see page 246

2-in glass fiber blanket

Free standing wall

Airspace as specified

1/4-in continuous joint packed with loose glass fiber and caulked with nonhardening mastic

For floor joint see detail page 241

FIG. 6.7 Typical freestanding wall isolated at top.

6.8 Structural Separation between Double Walls

In cases where extraordinarily low noise levels are required, such as concert halls, music buildings, conference centers, art and cultural projects, structural breaks may be needed between building spaces.

In Figure 6.8 such a complete structural separation was designed and built between the main mechanical equipment room and the rest of the building, in this case, a concert hall. To the left is the poured concrete wall of the concert hall, and on a separate slab to the right is the concrete masonry unit wall of the mechanical equipment room. The break at the slab, which has temporarily been covered with job debris (see arrow), must be thoroughly cleaned prior to final acceptance of the project.

Styrofoam board has been placed in the cavity between the walls and the slab separation. Instruction was given to the installing contractor, by the job supervisor, to clean off all cement grout at the styrofoam joints to eliminate bridging between the two walls.

FIG. 6.8 **Structural separation between double walls.**

6.9 Utilities in Wall

Figure 6.9 shows the intersection of several utilities in a wall that separates a noisy area from a noise-sensitive space. Careful planning and forethought could have prevented this arrangement. In this installation, the difficulties of achieving a proper closure are compounded, and the likelihood of effective noise control is reduced. See Figure 3.74 for the recommended method of installation.

FIG. 6.9 Utilities in wall.

FLOORS

6.10 **Floated Floors** ———————————————————————————————

6.11
6.12 Floated or resiliently supported secondary floors may be used when additional transmission-loss capability is required to control airborne noise transmission. For example, a 6-in-thick solid concrete slab weighing 70 lb/ft² is in a sound transmission class (STC) of STC 54. Doubling the thickness of the slab to 12 in theoretically increases the transmission loss to STC 59 or 60. This example shows that increasing the weight in order to gain a few decibels of STC soon becomes structurally impractical and uneconomical.

A composite floor, ceiling, wall, or partition system can achieve a much higher sound transmission loss than a solid one, by separating the noise or vibratory source from the supporting structures. The illustrations show glass fiber, sound-absorption material placed between isolators, but in many applications the fiberglass infill is omitted. Resilient isolators are strategically placed along the grid of centerlines; the isolators may be furnished precemented to the panels as shown in Figure 6.12, or they may be placed on the subfloor on the jobsite and covered with plywood as shown in Figures 6.10 and 6.11.

For the reduction of downwardly transmitted noise, it is normal practice to pour the reinforced concrete floating floor on isolation panels using water-packed plywood preset form. Figure 6.15 shows the elements of construction for the floor. Starting from the top and moving down, they are (1) the poured reinforced concrete floated floor, (2) the waterproof plywood form, (3) the resilient mounts and the fiberglass insulation, and (4) the structural slab. In addition, a resiliently hung gypsum board or plaster ceiling is shown below the structural slab. This may or may not be required depending on the amount of noise reduction needed.

FIG. 6.10 **Plywood with holes for inserting vibration isolators.**

FIG. 6.11 Vibration isolators being inserted.

FIG. 6.12 Fiberglass and plywood combination with isolators already place.

6.14   Resilient isolators for floated floors may be precom-
6.15   pressed fiberglass, neoprene, or steel spring, or in some
cases, combinations of these types. A typical precompressed fiberglass isolator is shown in Figure 6.13. A
spring in series with a neoprene pad is seen in Figure
6.14. This is a jack-up mount with a steel spring that may
be used in installations where input frequency is less
than 10 Hz or where frequencies less than 30 Hz having
dynamic amplitudes are expected. Where dynamic
frequency does not exceed 10 Hz, the neoprene jack-up
mounts shown in Figures 6.16 and 6.17 should be used.

    This jack-up isolator should be positioned so that it
picks up reinforcing rods in the two side brackets
specifically designed for this purpose (see Figure 6.15).

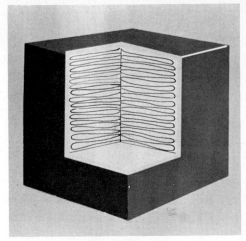

FIG. 6.13 **Fiberglass isolator with protective
coating.** *(Peabody Noise Control.)*

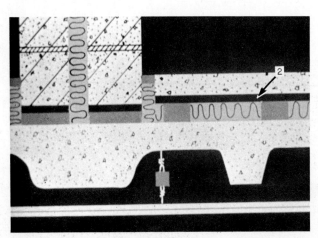

FIG. 6.15 **Typical floated-floor construction.** *(Peabody Noise
Control.)*

FIG. 6.14 **Jack-up floor isolator.** *(Peabody Noise
Control.)*

6.16 Jack-Up Isolators _____

6.17 Another type of isolator used for floating floors eliminates
the necessity for plywood. It is the jack-up isolator that
may contain either neoprene or steel springs.

Figures 6.16 and 6.17 show this type of isolator with
neoprene as the isolation material. These isolators are
placed on the floor over plastic sheeting, which prevents
bonding of the floating floor to the substructure. When
steel bars are used, they are coordinated with the steel
reinforcing, and when mesh is adequate, the isolators are
located in the mesh squares. When such a floating floor
has been poured and properly cured, the jack-up mounts
are adjusted to lift the floor to the desired height, thus
floating the floor as a totally separate unit. Fiberglass
infill is not required when this method is used.

FIG. 6.17 Individual floated-floor isolator with
reinforcing steel.

FIG. 6.16 Reinforcing steel and isolators for floated floor.

6.18 Resilient Structural Separations

6.19
6.20 A method for isolating slabs on grade or nonfloated floors is shown in Figure 6.18. Recommended methods for resiliently treating penetrations of floated floors by various utilities are shown in Figure 6.19, and fitting for floor and roof drains is shown in Figure 6.20.

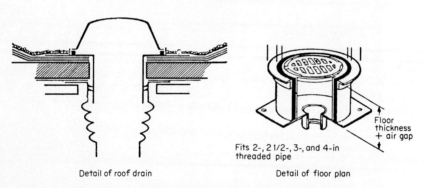

Detail of roof drain

Fits 2-, 2 1/2-, 3-, and 4-in threaded pipe

Floor thickness + air gap

Detail of floor plan

FIG. 6.20 Resilient fittings for floor and roof drains. *(Johns/Manville.)*

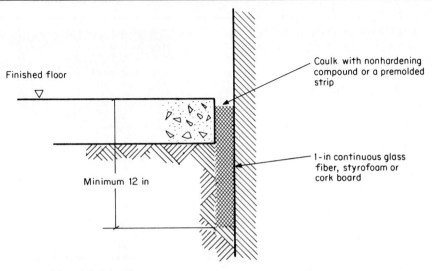

FIG. 6.18 Floor joint detail.

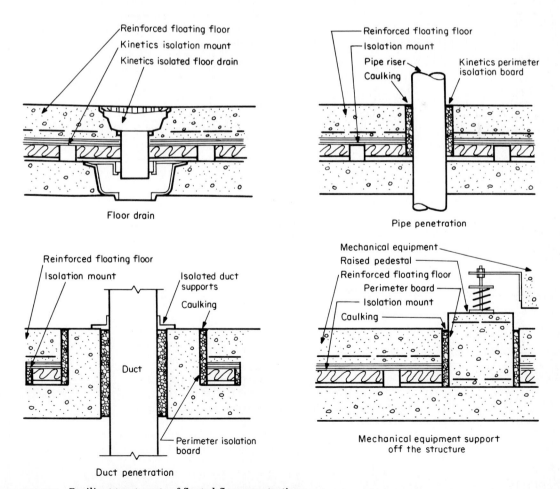

FIG. 6.19 Resilient treatments of floated-floor penetrations.

CEILINGS

6.21 Duct Connection in Ceiling

6.22 Figure 6.21 shows an approved method for connecting sheet-metal trunk duct to diffusers, grilles, or registers in a resiliently hung plasterboard or gypsum-board ceiling (see Figure 6.22). The branch duct from the trunk to the ceiling air terminal is made of fiberglass with aluminum foil facing. This duct is taped and sealed airtight to the trunk duct with approved, pressure-sensitive duct tape. The connection between the fiberglass duct section and the air-terminal device should be similarly treated, by removing the diffuser, grille, or register and taping and sealing the inside of the duct section to the neck of the air-terminal device.

FIG. 6.21   **Duct connection resiliently hung ceiling.**

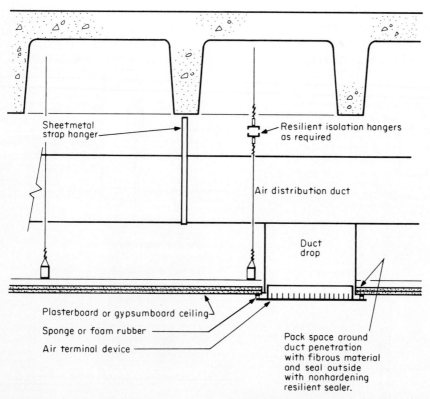

FIG. 6.22   **Typical detail of the duct penetration of plasterboard or gypsum-board ceilings.**

6.23 Resiliently Suspended Ceiling Framing

The type of vibration isolator shown in Figure 6.23 is effective for resiliently suspended gypsum-board or plaster ceilings, like those used in music practice rooms, auditoriums, or other noise-sensitive spaces. The vibration isolator has a rather soft (30-durometer) neoprene element in a metal housing. The fiberglass in the ceiling plenum created by the slab or building construction and the suspended ceiling is used to absorb some of the sound energy that may travel from the areas above or below the plenum. In this picture, the fiberglass is supported by chicken-wire mesh.

FIG. 6.23   Resiliently suspended ceiling framing.

6.24 Resiliently Suspended Ceiling with Sway Bracing

The ceiling treatment in Figure 6.24 is similar to the one in Figure 6.23 except that resilient sway bracing has been applied to restrict the movement of the large resiliently suspended ceiling sections in seismic zones. The isolators used for the ceiling hangers and the sway braces are of the same type.

FIG. 6.24   Resiliently suspended ceiling with resilient sway bracing.

243

6.26
6.27
6.28
6.29
Many types of ceiling installations may be hung resiliently by the use of conventional vibration-isolation hangers. These hangers may have a spring in series with a neoprene element like those in Figures 6.25 and 6.26. Additional details for using resiliently hung ceilings with resiliently mounted walls are shown in Figures 6.28 and 6.29. Note that the hangers are attached at the top to wood blocking pieces securely nailed in between the joists. The bottom hanger wire is attached to 2 × 4 horizontal runners, and then the sheetrock or gypsum board is secured directly to the bottom of the 2 × 4 runners (see Figure 6.26). Hanger spacing should be arranged so that each hanger is properly loaded to achieve the desired deflection and isolation. Of course, no construction or utilities should be allowed to form a rigid connection or bridge between the resiliently suspended ceiling and the construction above. Neither should the suspended ceiling touch the walls of the room. See Figure 6.27 for recommended acoustical treatment at these points.

Other types of vibration-isolation hangers are also available and should be selected on the basis of weight-loading and static deflection requirements.

FIG. 6.25   Resiliently suspended ceiling support member.

FIG. 6.26   Resiliently suspended ceiling supports.

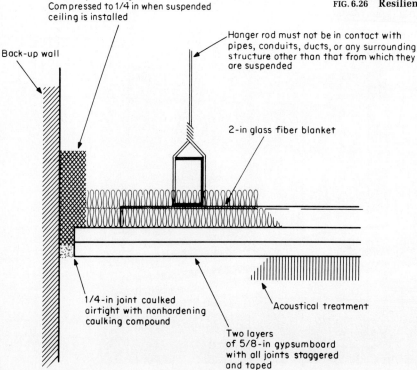

1-in glass fiber blanket fastened to periphery of ceiling structure. Compressed to 1/4 in when suspended ceiling is installed

Back-up wall

Hanger rod must not be in contact with pipes, conduits, ducts, or any surrounding structure other than that from which they are suspended

2-in glass fiber blanket

1/4-in joint caulked airtight with nonhardening caulking compound

Acoustical treatment

Two layers of 5/8-in gypsumboard with all joints staggered and taped

FIG. 6.27   Typical suspended ceiling joint detail.

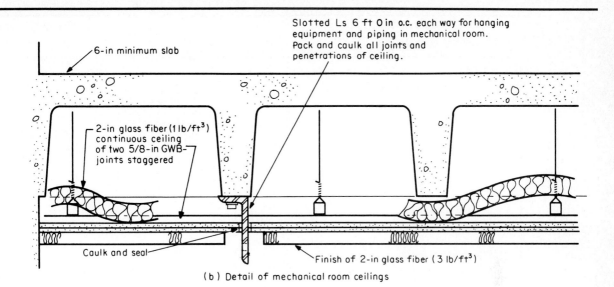

Slotted Ls 6 ft 0 in o.c. each way for hanging
equipment and piping in mechanical room.
Pack and caulk all joints and
penetrations of ceiling.

6-in minimum slab

2-in glass fiber (1 lb/ft^3)
continuous ceiling
of two 5/8-in GWB—
joints staggered

Caulk and seal

Finish of 2-in glass fiber (3 lb/ft^3)

(b) Detail of mechanical room ceilings

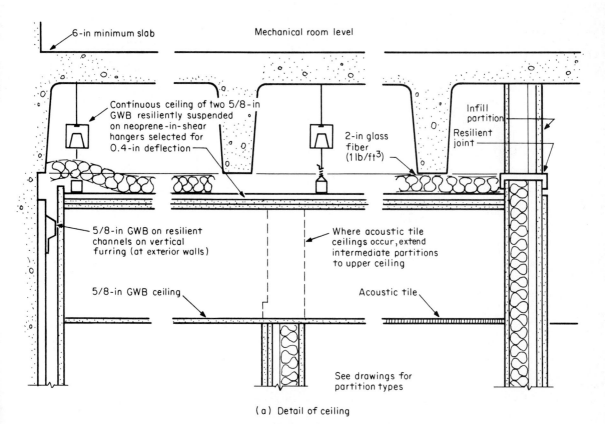

6-in minimum slab

Mechanical room level

Continuous ceiling of two 5/8-in
GWB resiliently suspended
on neoprene-in-shear
hangers selected for
0.4-in deflection

2-in glass
fiber
(1 lb/ft^3)

Infill
partition

Resilient
joint

5/8-in GWB on resilient
channels on vertical
furring (at exterior walls)

Where acoustic tile
ceilings occur, extend
intermediate partitions
to upper ceiling

5/8-in GWB ceiling

Acoustic tile

See drawings for
partition types

(a) Detail of ceiling

FIG. 6.28 Recommended resilient mountings for walls and ceilings for mechanical equipment rooms.

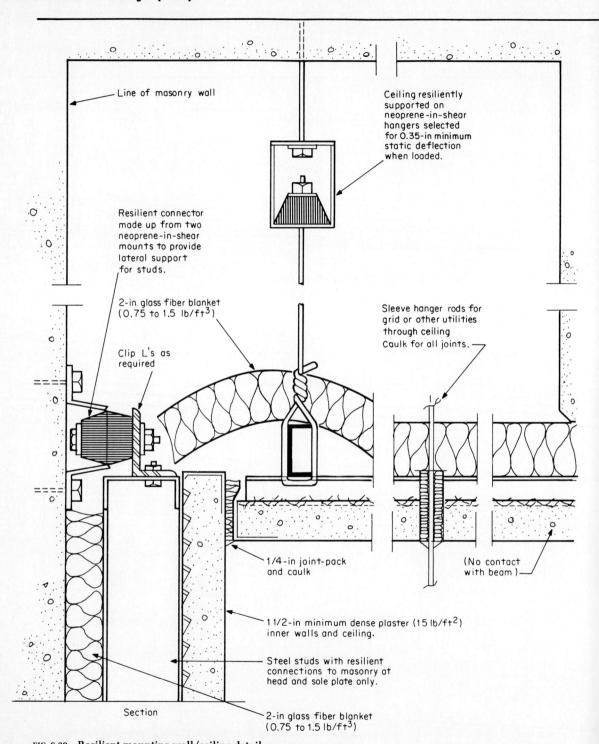

Line of masonry wall

Ceiling resiliently
supported on
neoprene-in-shear
hangers selected
for 0.35-in minimum
static deflection
when loaded.

Resilient connector
made up from two
neoprene-in-shear
mounts to provide
lateral support
for studs.

2-in glass fiber blanket
(0.75 to 1.5 lb/ft^3)

Clip L's as
required

Sleeve hanger rods for
grid or other utilities
through ceiling
Caulk for all joints.

1/4-in joint-pack
and caulk

(No contact
with beam)

1 1/2-in minimum dense plaster (15 lb/ft^2)
inner walls and ceiling.

Steel studs with resilient
connections to masonry at
head and sole plate only.

Section

2-in glass fiber blanket
(0.75 to 1.5 lb/ft^3)

FIG. 6.29   Resilient mounting wall/ceiling detail.

6.30 Inertia Hangers for Ceilings

6.31
6.32

Inertia vibration-isolation hangers such as those shown in Figure 6.30 may be used at times when the additional weight provided by the mass of the isolator can be used effectively. For instance, a lightweight ceiling may be responding to noise and vibration the way a drum head responds when struck. Lightweight ceilings are like diaphragms. When the vibratory energy comes down from the supporting structure, rather than the static mass assigned to each isolator acting as one complete mass, only a small portion of that mass is set into motion by the energy coming down the spring. Thus, there can be considerable excitation. By interposing the concentrated mass of the hanger before this energy reaches the ceiling, we establish what might be called a quiet point for ceiling attachment, and in this way the extra weight tends to dampen the excitation coming from the structure to the ceiling.

Another case in which inertia hangers may be used occurs when a very light ceiling does not have enough weight in itself to deflect the hanger isolators properly. The added mass of an inertia unit may then be introduced to load the isolators effectively.

Like all isolators, inertia hangers must be installed and maintained in correct position to function as intended. One way of installing inertia hangers is shown in Figure 6.31. A close-up of one such hanger, seen in Figure 6.32, reveals, however, that the effectiveness of these isolators may be considerably reduced if they come into direct contact with other utilities. In this case, the other utility is a section of rigid electrical conduit running to a junction box. The box, in turn, is securely fastened to the building.

FIG. 6.30 Inertia isolator.

FIG. 6.31 Inertia isolator installed.

The added weight introduced by the use of inertia hangers could be a problem if they are used in an earthquake zone. Thus, if no other, more practical method can be used to solve the isolation problem, extra care should be taken in such instances to provide seismic protection for the hanger.

FIG. 6.32   **Several inertia isolators installed.**

DOORS

6.33 Door Gaskets

6.34
6.35

In many instances, it becomes necessary to use specially constructed doors that are certified to have a rated sound transmission class (STC).

In walls that have been constructed to have a given STC rating, doors often represent the weak link, acoustically. To ensure the maximum STC of a door and its installation in a rated wall, good door gaskets must be included so that when the door is closed, it forms an airtight seal.

Several types and application methods for satisfactory door gaskets are shown on Figures 6.33 to 6.35. If optimum results are to be achieved from a noise-transmission point of view, both the door and its jamb and threshold must be plumb and neatly fitted so that the door gaskets will work. Even in the best installations, minor adjustments may be required from time to time to ensure the continued tight seal all around the door.

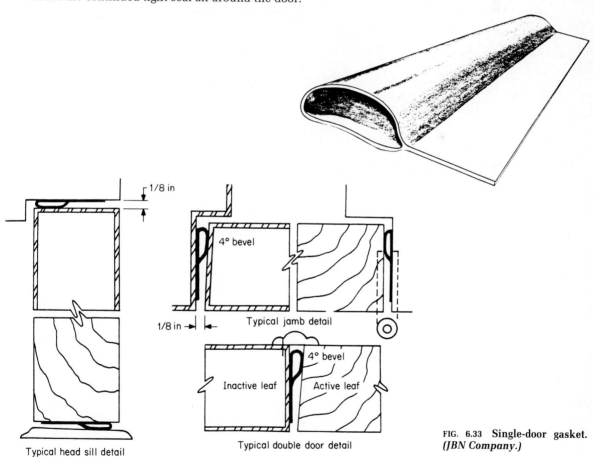

FIG. 6.33 **Single-door gasket.** *(JBN Company.)*

1/8 in

4° bevel

Typical jamb detail

1/8 in

Typical head sill detail

Inactive leaf

Active leaf

4° bevel

Typical double door detail

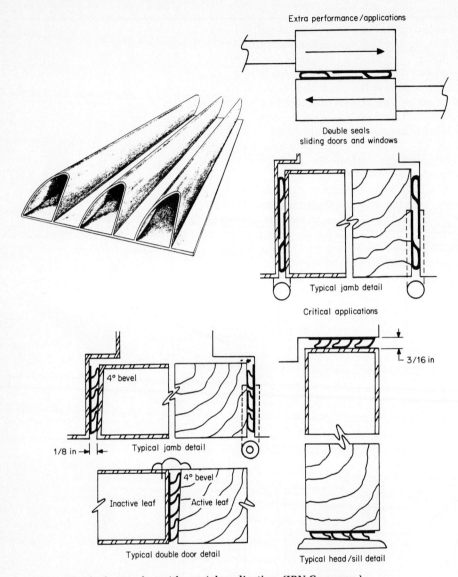

Extra performance /applications

Double seals
sliding doors and windows

Typical jamb detail

Critical applications

4° bevel

1/8 in ──►|◄─   Typical jamb detail

3/16 in

4° bevel

Inactive leaf   Active leaf

Typical double door detail

Typical head /sill detail

FIG. 6.34   **Triple-door gasket with special application.** *(JBN Company.)*

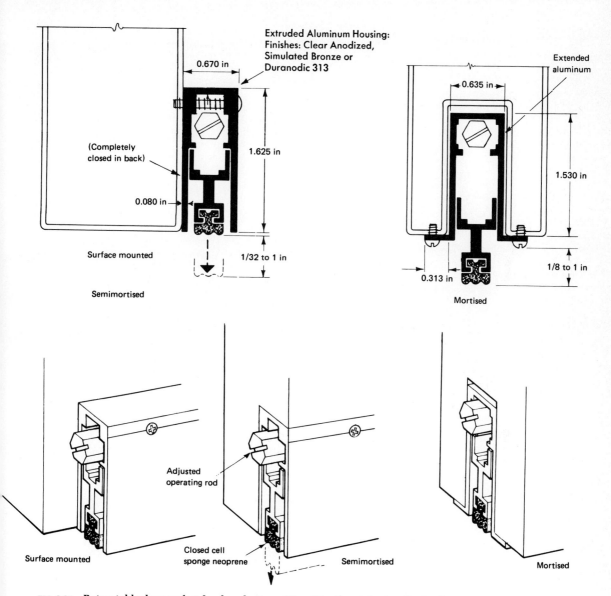

Extruded Aluminum Housing:
Finishes: Clear Anodized,
Simulated Bronze or
Duranodic 313

0.670 in

(Completely
closed in back)

1.625 in

0.080 in

Surface mounted

1/32 to 1 in

Semimortised

Extended
aluminum

0.635 in

1.530 in

0.313 in

1/8 to 1 in

Mortised

Surface mounted

Adjusted
operating rod

Closed cell
sponge neoprene

Semimortised

Mortised

FIG. 6.35 Retractable door gasket for door bottom. *(Zero Weatherstripping Co. Inc.)*

Seismic Isolation and Protection

SEISMIC PROTECTION OF RESILIENTLY MOUNTED SYSTEMS

As more land areas around the world are developed, more people are likely to experience a seismic event.* The protection of life and property from the devastating effects of earthquakes is already an urgent worldwide problem. There is a high probability of occurrence of earthquakes throughout the world in the next few years.† However, as natural phenomena, earthquakes usually do not kill people. The real lethal force during any seismic event is the movement and breaking up of man-made structures under the impact of the earthquake. This chapter, therefore, deals with the effects of earthquakes on man-made structures, and more specifically, the effects of earthquakes on resiliently isolated mechanical equipment in buildings.

One important difference distinguishes the seismic protection problems outlined in this chapter from the general equipment vibration problems described in the rest of this manual. A typical vibration problem can usually be worked on over a long period of time with many chances at a favorable solution; earthquake problems usually manifest themselves only when it is too late—after the seismic event has occurred. Much effort, therefore, must be put into advance planning and preinstallation assessment. The introduction to this chapter offers a discussion of some of the considerations that should be addressed before installing seismic protection. We then proceed to examine the utility of air springs and steel springs as vibration isolators and to delineate some of the specific problems with equipment in nuclear plants and with piping configurations. As a general aid to the reader, seismic snubber tests are outlined in Appendix C and Appendixes D, E, F, and G contain excerpts from earthquake regulations.

---

* *World Atlas of Seismic Zones and Nuclear Power Plants,* Wyle Laboratories, El Segundo, Calif., September 1979. See Appendix I for work maps of seismic zones.

† "Strong Motion Earthquake Instrument Arrays," *Proceedings of the International Workshop of Strong-Motion Earthquake Instrument Arrays,* May 2–5, 1978, Honolulu, Hawaii. Resolution, p. ix.

7.2 Scientists have a good understanding of the nature of earthquake ground motion and its effect on buildings and other structures, such as rocking and twisting. Basically, only four types of elastic waves are responsible for the shaking that is felt and the resultant damage caused by an earthquake. These are the P wave, the S wave, the Love wave, and the Raleigh wave. These waves are similar in many important aspects to sound waves in air, e.g., in their method of travel through both solids and liquids as well as in their analysis.

Two general points about the power of earthquakes and seismic protection should be made at the beginning of this discussion. First, the force released and transmitted during a seismic event can be tremendous. An

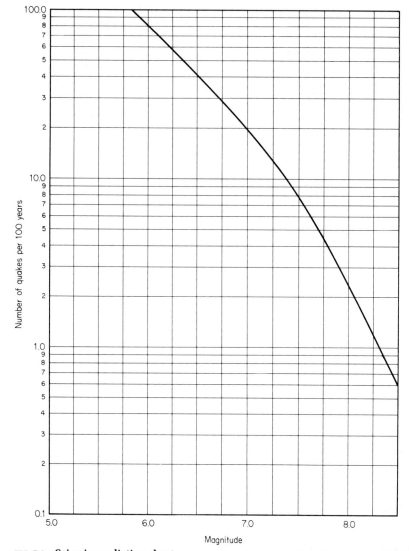

FIG. 7.1 Seismic prediction chart.

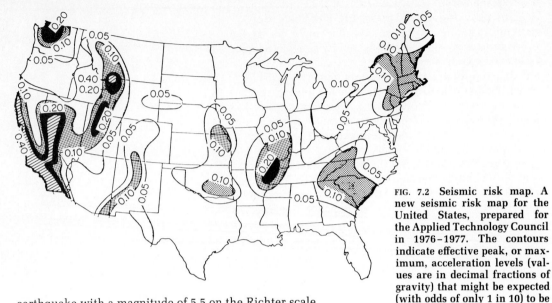

FIG. 7.2 Seismic risk map. A new seismic risk map for the United States, prepared for the Applied Technology Council in 1976–1977. The contours indicate effective peak, or maximum, acceleration levels (values are in decimal fractions of gravity) that might be expected (with odds of only 1 in 10) to be exceeded during a 50-year period. *(From B. A. Bolt, Earthquakes—A Primer, W. H. Freeman, San Francisco, 1978.)*

earthquake with a magnitude of 5.5 on the Richter scale turns out to have an energy level of about 10^{20} ergs. By way of comparison, the energy of the Bikini nuclear blast in 1946 was 10^{19} ergs.* A measure of this force is how large an area can be severely affected by a major earthquake. For example, the City of Los Angeles, which seismologists believe can experience up to 6 min of continuous strong ground shaking in the case of a major seismic event on the San Andreas fault, is more than 30 mi from the fault.

However, it is important to stress a second point; in designing and installing earthquake protection, the purpose is not to protect for major or massive earthquakes (7 or higher on the Richter scale). Seismic events of this magnitude will cause most high-rise structures to sustain so much damage that they may require demolition and replacement.

Instead, the intent of seismic protection is to keep mechanical equipment and systems operational, or at least in place, during minor and moderate earthquakes that cause reparable structural damage. Minor to moderate earthquakes are much more prevalent than major ones during any given period of time, and therefore, during their useful life span, many structures will experience one or more minor earthquakes but not a major seismic event. A good example would be the seismic prediction chart shown in Fig. 7.1 for the State of California; it can be noted that even a Richter 6 (moderate earthquake) will occur somewhere in California on the average of every 2 years. Also see the seismic risk map of the United States (Figure 7.2).

---

* B. A. Bolt, *Earthquakes—A Primer*, W. H. Freeman, San Francisco, 1978, p. 108.

EQUIPMENT MOUNTING FOR SEISMIC PROTECTION

The design of optimal seismic protection will depend on a number of judgments by the engineer who will have to decide, for instance, if equipment is to be kept in place with no regard for continued operation or breakup or if the equipment is to be kept in place with minor damage and is to be operational. This decision, in turn, will rest on judging how essential the equipment is.*

A prime piece of equipment, such as a main air supply system or a pump in a vital water supply system, could be essential for life support in the building; therefore, its operation after a seismic event is essential. On the other hand, a toilet exhaust fan is not essential to operate the building in emergency, but it must be kept in place to avoid its causing damage to life or property. Its continued operation after a seismic event is not essential for a time.

Fundamental judgments, such as the preceding, will help in selecting the type of equipment mounting and the kind of seismic isolation to be used. Generally, equipment can either be hard-mounted, that is, bolted to the building, or resiliently mounted.

When equipment is hard-bolted or secured to the building, the forces that must be controlled are much less. Such equipment can withstand the effects of an earthquake better than resiliently mounted equipment, because it moves with the structure and does not experience a seismic force increase across the snubber.

Therefore, equipment such as electric generators or fire pumps, which operate only in emergencies, should be hard-mounted if possible. Hard-mounted equipment is simple and straightforward to install. Because the structural engineer is equipped to handle the details, these installations are not of primary concern here.

Resiliently mounted equipment, however, is not structurally part of the building system. Unless such equipment is seismically restrained, one of the earthquake's frequencies may resonate the mounting's system and cause excessive amplitudes, causing the equipment to fall or fly off its resilient supports. The installations described in this chapter illustrate the kinds of seismic snubbing that can be used with resiliently mounted equipment and the problems encountered by each installation.

Devices used to provide earthquake vibration restraints are commonly called *seismic restraints* or *snubbers*. A seismic snubber or restraint is a device that does not interfere with the vibration-isolator system under

---

* "Study to Establish Seismic Protection Provisions for Furniture, Equipment & Supplies for VA Hospitals," Research Staff Office of Construction, Washington, D.C., January 1976.

normal operating conditions; it is engaged only in the event of a seismic input. Because of the built-in air gap, snubbers tend to increase the seismic forces. This occurs because at some point in the earthquake cycle, the structure holding the outside of the snubber may be moving in a direction opposite to that of the equipment. In simplified terms, at the time of contact, we are dealing with a double velocity. Despite the cushioning of the snubber, which allows for deceleration, the ultimate forces are larger than if the equipment were rigidly bolted to the structure.

It is essential to remember that none of the mechanical systems in buildings are wholly self-contained. External connections such as electric conduits, water piping, and sheet-metal duct connections must be able to withstand even the limited movements allowed by a relatively sophisticated seismic snubbing system.

To choose the right seismic restraint, one might use either static analysis or dynamic analysis.* For the first time, seismic snubber selection guide charts are included in this manual to help in selecting the proper snubbing device for various seismic zones† and building types (see Figures 7.3 to 7.6). These charts are only a guide; the author cannot be responsible for their use.

If equipment is to be kept in place with no regard for continued operation or breakdown, a static analysis that satisfies local codes is often used. G levels of the restraints are called out by the code and the restraints and their attachments as well as the supporting structure are designed at least to these levels. Since the equipment moves against these restraints, with a hammering effect, the shock to the equipment may render it nonoperational, and in fact, the dynamic forces may be large enough to break through the restraints that were designed statically. Whether the equipment can continue in operation would depend on whether the impact exceeds the equipment's fragility level.

If equipment is to be both kept in place and operational, a dynamic analysis should be used, although a dynamic analysis is a good deal more complex than the static calculations. As with all sophisticated, accurately designed systems, dynamic snubbers with their smaller clearances are somewhat more difficult to install, as explained in this chapter.

─────────────────

* See *Seismic Analysis*, Wyle Laboratories, El Segundo, Calif., and Bulletin SCS-100, Mason Industries, Happauge, N.Y.

† See *Seismic Risk and Susceptibility Zones in the U.S. World Atlas*, by Wyle Laboratories, El Segundo, Calif., and for zones in the rest of the world, see Appendix I.

FIG. 7.3 **Seismic snubber and analysis selection chart for hospital structures in zone 4.**

Equipment Importance Factor	Primary: Must Remain Operational during and after Earthquake			Secondary: Can Be Restored to Operation within 48 h after Earthquake			Nonessential: Must Be Kept in Place		
Equipment Location Elevation	Up to 25 ft	26–110 ft	111 ft and Up	Up to 25 ft	26–110 ft	111 ft and Up	Up to 25 ft	26–110 ft	111 ft and Up
Refrigeration machines									
Absorption	AJ	BGI	BGI	AJ	AJ	BGI			
Centrifugal chiller or heat pumps	AJ	BGI	BGI	AJ	AJ	BGI			
Cooler condensed mounted hermetic compressors	AJ	BGI	BGI	AJ	AJ	BGI			
Cooler condenser along side hermetic compressor	AJ	BGI	BGI	AJ	AJ	BGI		Always	
Reciprocating compressors									
Up to 4000 lb	AJ	AGI	BGI	CJ	CJ	BJ		primary	
Over 4000 lb	AJ	BGI	BGI	AJ	AJ	BGI		or	
Reciprocating chillers or heat pumps								secondary	
Up to 4000 lb	AJ	AGI	BGI	CJ	CJ	BJ			
Over 4000 lb	AJ	BGI	BGI	AJ	AJ	BGI			
Averaged steam generators (boilers)	AJ	AGJ	AGI	AJ	AGJ	AGJ	AGJ	AGJ	AGJ
Pumps									
Close-coupled									
Through 5 hp	AJ	AJ	AGJ	AJ	AJ	AJ	A	A	AJ
7.5 hp and larger	AJ	AGJ	AGJ	AJ	AJ	AGJ	A	A	AJ
Base-mounted									
Up to 60 hp	AJ	BGJ	BGI	AGJ	AGJ	AGI	AJ	AJ	AJ
75 hp and larger	AGJ	BGJ	BGI	AGJ	BGJ	BGI	AJ	AJ	AJ
Factory-assembled H and V units									
Curb-mounted rooftop units	F	F	FJ	F	F	FJ	F	F	FJ
Suspended units									
Through 5 hp	FJ	FJ	FJ	F	FJ	FJ	F	F	FJ
7.5 hp and larger	FJ	FJ	FI	F	FJ	FJ	F	FJ	FJ

Note: In the absence of the letters I or J, no analysis is required, but snubbers must be selected for g forces shown in charts. Type C may always be substituted for type A to facilitate installation.

Equipment Importance Factor	Primary: Must Remain Operational during and after Earthquake			Secondary: Can Be Restored to Operation within 48 h after Earthquake			Nonessential: Must Be Kept in Place		
Equipment Location Elevation	Up to 25 ft	26–110 ft	111 ft and Up	Up to 25 ft	26–110 ft	111 ft and Up	Up to 25 ft	26–110 ft	111 ft and Up
Factory-assembled H and V units (Cont.)									
Floor-mounted units									
Through 5 hp	C	CJ	CJ	C	CJ	CJ	C	C	CJ
7.5–40 hp	C	AJ	AI	C	CJ	AJ	C	CJ	CJ
50 hp and larger	CJ	AI	BI	CJ	AJ	BJ	A	AJ	AJ
Air compressors									
Tank type	A	BJ	BI	A	AJ	AJ	A	AJ	AJ
V-W type	A	BJ	BI	A	AJ	AJ	A	AJ	AJ
Horizontal-vertical, 1 or 2 cylinders									
275–499 r/min	AJ	BJ	BI	A	AJ	BJ	A	AJ	AJ
500–800 r/min	AJ	BJ	BI	A	AJ	BJ	A	AJ	AJ
Blowers									
Utility sets									
floor-mounted	C	CJ	CJ	C	C	CJ	C	C	C
Suspended centrifugal type	CH	CHJ	CHI	CH	CH	CHJ	CH	CH	CH
Fan heads									
floor-mounted	A	AJ	AI	A	AJ	AJ	A	A	AJ
Suspended tubular centrifugal and axial fans									
Suspended up to									
25 hp	F	F	FJ	F	F	FJ	F	F	F
30 hp and larger	F	FJ	FI	F	FJ	FI	F	F	FJ
Floor-mounted motor									
on/in any fan casing	A	AJ	AJ	A	AJ	AJ	A	A	AJ
Floor-mounted arrangement, 1 or any separately mounted motor	A	AJ	AJ	A	AJ	AJ	A	A	AJ
Cooling towers and condensing units	AJ	BJ	BI	AJ	BJ	BI	AJ	AJ	AJ
Isolated pipe 6-in diameter and smaller	F	F	FJ	F	F	FJ	F	F	F
Isolated pipe 8-in diameter and larger	FJ	FJ	FI	FJ	FJ	FJ	F	FJ	FJ
Risers 6-in diameter and larger in excess of 160 ft	EJ	EJ	EJ	EJ	EJ	EJ	Always essential		

FIG. 7.4 **Seismic snubber and analysis selection chart for commercial high-rise structures in zone 4.**

Equipment Importance Factor	Primary: Must Remain Operational during and after Earthquake			Secondary: Can Be Restored to Operation within 48 h after Earthquake			Nonessential: Must Be Kept in Place		
Equipment Location Elevation	Up to 25 ft	26–110 ft	111 ft and Up	Up to 25 ft	26–110 ft	111 ft and Up	Up to 25 ft	26–110 ft	111 ft and Up
Refrigeration machines									
Absorption	AJ	BJ	BGI	A	AJ	BGJ			
Centrifugal chiller or heat pumps	AJ	BJ	BGI	A	AJ	BGJ			
Cooler condensed mounted hermetic compressors	AJ	BJ	BGI	A	AJ	BGJ		Always	
Cooler condenser along side hermetic compressor	AJ	BJ	BGI	A	AJ	BGJ		primary	
Reciprocating compressors								or	
Up to 4000 lb	A	AJ	BGJ	A	CJ	BGJ			
Over 4000 lb	AJ	BJ	BGI	AJ	AJ	BGJ		secondary	
Reciprocating chillers or heat pumps									
Up to 4000 lb	A	AJ	BGJ	A	AJ	BGJ			
Over 4000 lb	AJ	BJ	BGI	AJ	AJ	BGJ			
Packaged steam generators (boilers)	AJ	AGJ	AGJ	AJ	AJ	AGJ	AJ	AJ	AJ
Pumps									
Close-coupled Through									
5 hp	AJ	AJ	AJ	A	A	AJ	A	A	A
7.5 hp and larger	A	AJ	AGJ	A	A	AJ	A	A	AJ
Base-mounted									
Up to 60 hp	AJ	AGJ	BGJ	AJ	AJ	AGJ	AJ	AJ	AJ
75 hp and larger	AJ	AGJ	BGI	AJ	AJ	AGJ	AJ	AJ	AJ
Factory-assembled H and V units									
Curb-mounted rooftop units	F	F	FJ	F	F	FJ	F	F	FJ
Suspended units Through									
5 hp	F	FJ	FJ	F	F	FJ	F	F	F
7.5 hp and larger	F	FJ	FI	F	FJ	FJ	F	F	FJ

Note: In the absence of the letters I or J, no analysis is required, but snubbers must be selected for g forces shown in charts. Type C may always be substituted for type A to facilitate installation.

Equipment Importance Factor	Primary: Must Remain Operational during and after Earthquake			Secondary: Can Be Restored to Operation within 48 h after Earthquake			Nonessential: Must Be Kept in Place		
Equipment Location Elevation	Up to 25 ft	26–110 ft	111 ft and Up	Up to 25 ft	26–110 ft	111 ft and Up	Up to 25 ft	26–110 ft	111 ft and Up
Factory-assembled H and V units (Cont.)									
Floor-mounted units									
Through 5 hp	C	CJ	CJ	C	C	CJ	C	C	C
7.5–40 hp	C	CJ	AI	C	CJ	AJ	C	C	CJ
50 hp and larger	CJ	AJ	BI	CJ	AJ	BJ	A	A	AJ
Air compressors									
Tank type	A	AJ	AI	A	AJ	AJ	A	A	AJ
V-W type	A	AJ	BI	A	AJ	BJ	A	AJ	AJ
Horizontal-vertical, 1 or 2 cylinders									
275–499 r/min	AJ	BJ	BJ	AJ	BJ	BJ	A	AJ	AJ
500–800 r/min	AJ	BJ	BJ	BJ	BJ	BJ	A	AJ	AJ
Blowers									
Utility sets									
floor-mounted	C	CJ	CJ	C	C	CJ	C	C	C
Suspended centrifugal type	CH	CH	CHJ	CH	CH	CHJ	CH	CH	CH
Fan heads									
floor-mounted	A	AJ	AJ	A	AJ	AJ	A	A	AJ
Suspended tubular centrifugal and axial fans									
Suspended up to 25 hp	F	F	FJ	F	F	FJ	F	F	F
30 hp and larger	F	FJ	FI	F	FJ	FI	F	F	FJ
Floor-mounted-motor on/in fan casing	A	AJ	AJ	A	AJ	AJ	A	A	AJ
Floor-mounted arrangement, 1 or any separately mounted motor	A	AJ	AJ	A	AJ	AJ	A	A	AJ
Cooling towers and condensing units	AJ	AJ	BI	AJ	AJ	BJ	AJ	AJ	AJ
Isolated pipe 6-in diameter and smaller	F	F	FJ	F	F	FJ	F	F	F
Isolated pipe 8-in diameter and larger	FJ	FJ	FI	FJ	FJ	FJ	F	F	FJ
Risers 6-in diameter and larger in excess of 160 ft	EJ	EJ	EJ	EJ	EJ	EJ	Always Essential		

FIG. 7.5 **Seismic snubber and analysis selection chart for hospitals and commercial high-rise structures in zone 3.**

Equipment Importance Factor	Primary: Must Remain Operational during and after Earthquake			Secondary: Can Be Restored to Operation within 48 h after Earthquake			Nonessential: Must Be Kept in Place		
Equipment Location Elevation	Up to 25 ft	26–110 ft	111 ft and Up	Up to 25 ft	26–110 ft	111 ft and Up	Up to 25 ft	26–110 ft	111 ft and Up
Refrigeration machines									
Absorption	AJ	AJ	BGI	AJ	BJ	BGJ			
Centrifugal chiller or heat pumps	AJ	AJ	BGI	AJ	BJ	BGJ			
Cooler condensed mounted hermetic compressors	AJ	AJ	BGI	AJ	BJ	BGJ		Always	
Cooler condenser along side hermetic compressor	AJ	AJ	BGI	AJ	BJ	BGJ		primary	
Reciprocating compressors									
Up to 4000 lb	A	AJ	BGJ	C	CJ	BGJ		or	
Over 4000 lb	AJ	BJ	BGI	AJ	AJ	BGJ		secondary	
Reciprocating chillers or heat pumps									
Up to 4000 lb	A	AJ	BGJ	A	CJ	BGJ			
Over 4000 lb	AJ	BJ	BGI	AJ	AJ	BGJ			
Packaged steam generators (boilers)	AJ	AGJ	AGJ	AJ	AGJ	AGJ	A	AJ	AJ
Pumps									
Close-coupled									
Through 5 hp	A	A	AJ	A	A	A	A	A	A
7.5 hp and larger	A	A	AJ	A	A	AJ	A	A	A
Base-mounted									
Up to 60 hp	A	AGJ	AGJ	A	AJ	AGJ	A	AJ	AJ
75 hp and larger	A	AGJ	BGI	A	AJ	AGJ	A	AJ	AJ
Factory-assembled H and V units									
Curb-mounted rooftop units	F	F	FJ	F	F	FJ	F	F	FJ
Suspended units Through 5 hp	F	F	FJ	F	F	FJ	F	F	F
7.5 hp and larger	F	FJ	FJ	F	F	FJ	F	F	F

Note: In the absence of the letters I or J, no analysis is required, but snubbers must be selected for g forces shown in charts. Type C may always be substituted for type A to facilitate installation.

Equipment Importance Factor / Equipment Location Elevation	Primary: Must Remain Operational during and after Earthquake			Secondary: Can Be Restored to Operation within 48 h after Earthquake			Nonessential: Must Be Kept in Place		
	Up to 25 ft	26–110 ft	111 ft and Up	Up to 25 ft	26–110 ft	111 ft and Up	Up to 25 ft	26–110 ft	111 ft and Up
Factory-assembled H and V units (Cont.)									
Floor-mounted units									
Through 5 hp	C	CJ	CJ	C	C	CJ	C	C	C
7.5–40 hp	C	CJ	AI	C	CJ	AJ	C	C	CJ
50 hp and larger	CJ	AJ	BI	CJ	AJ	BJ	A	A	AJ
Air compressors									
Tank type	A	AJ	AI	A	AJ	AJ	A	A	AJ
V-W type	A	AJ	BI	A	AJ	BJ	A	AJ	AJ
Horizontal-vertical, 1 or 2 cylinders									
275–499 r/min	AJ	BJ	BJ	AJ	BJ	BJ	A	AJ	AJ
500–800 r/min	AJ	BJ	BJ	BJ	BJ	BJ	A	AJ	AJ
Blowers									
Utility sets									
floor-mounted	C	CJ	CJ	C	C	CJ	C	C	C
Suspended centrifugal type	CH	CH	CHJ	CH	CH	CHJ	CH	CH	CH
Fan heads									
floor-mounted	A	AJ	AJ	A	AJ	AJ	A	A	AJ
Suspended tubular centrifugal and axial fans									
Suspended up to									
25 hp	F	F	FJ	F	F	FJ	F	F	F
30 hp and larger	F	FJ	FI	F	FJ	FI	F	F	FJ
Floor-mounted-motor on/in fan casing	A	AJ	AJ	A	AJ	AJ	A	A	AJ
Floor-mounted arrangement, 1 or any separately mounted motor	A	AJ	AJ	A	AJ	AJ	A	A	AJ
Cooling towers and condensing units	AJ	AJ	BI	AJ	AJ	BJ	AJ	AJ	AJ
Isolated pipe 6-in diameter and smaller	F	F	FJ	F	F	FJ	F	F	F
Isolated pipe 8-in diameter and larger	FJ	FJ	FI	FJ	FJ	FJ	F	F	FJ
Risers 6-in diameter and larger in excess of 160 ft	EJ	EJ	EJ	EJ	EJ	EJ	Always Essential		

FIG. 7.6   **Seismic snubber and analysis selection chart for hospitals and commercial high-rise structures in zones 1 and 2.**

Equipment Importance Factor	Primary: Must Remain Operational during and after Earthquake			Secondary: Can Be Restored to Operation within 48 h after Earthquake			Nonessential: Must Be Kept in Place		
Equipment Location Elevation	Up to 25 ft	26–110 ft	111 ft and Up	Up to 25 ft	26–110 ft	111 ft and Up	Up to 25 ft	26–110 ft	111 ft and Up
Refrigeration machines									
Absorption	AJ	AJ	AJ	AJ	AJ	AJ			
Centrifugal chiller or heat pumps	AJ	AJ	AJ	AJ	AJ	AJ			
Cooler condensed mounted hermetic compressors	AJ	AJ	AJ	AJ	AJ	AJ			
Cooler condenser along side hermetic compressor	AJ	AJ	AJ	AJ	AJ	AJ	Always		
Reciprocating compressors							primary		
Up to 4000 lb	A	A	AJ	A	A	AJ			
Over 4000 lb	A	AJ	AJ	A	A	AJ	or		
Reciprocating chillers or heat pumps							secondary		
Up to 4000 lb	A	A	AJ	A	A	AJ			
Over 4000 lb	A	AJ	AJ	A	A	AJ			
Packaged steam generators (boilers)	AJ	AJ	AJ	AJ	AJ	AJ	A	A	AJ
Pumps									
Close-coupled through 5 hp	A	A	A	A	A	A	A	A	A
7.5 hp and larger	A	A	AJ	A	A	A	A	A	A
Close-mounted									
Up to 60 hp	A	A	AJ	A	A	AJ	A	A	A
75 hp and larger	A	A	AJ	A	A	AJ	A	A	A
Factory-assembled H and V units									
Curb-mounted rooftop units	F	F	F	F	F	F	F	F	F
Suspended units Through 5 hp	F	F	F	F	F	F	F	F	F
7.5 hp and larger	F	F	FJ	F	F	FJ	F	F	F

Note: In the absence of the letter I or J, no analysis is required, but snubbers must be selected for g forces shown in charts. Type C may always be substituted for type A to facilitate installation.

Equipment Importance Factor	Primary: Must Remain Operational during and after Earthquake			Secondary: Can Be Restored to Operation within 48 h after Earthquake			Nonessential: Must Be Kept in Place		
Equipment Location Elevation	Up to 25 ft	26–110 ft	111 ft and Up	Up to 25 ft	26–110 ft	111 ft and Up	Up to 25 ft	26–110 ft	111 ft and Up
Factory-assembled H and V units (Cont.)									
Floor-mounted units									
Through 5 hp	C	CJ	CJ	C	C	CJ	C	C	C
7.5–40 hp	C	CJ	AI	C	CJ	AJ	C	C	CJ
50 hp and larger	CJ	AJ	BI	CJ	AJ	BJ	A	A	AJ
Air compressors									
Tank type	A	AJ	AI	A	AJ	AJ	A	A	AJ
V-W type	A	AJ	BI	A	AJ	BJ	A	AJ	AJ
Horizontal-vertical, 1 or 2 cylinders									
275–499 r/min	AJ	BJ	BJ	AJ	BJ	BJ	A	AJ	AJ
500–800 r/min	AJ	BJ	BJ	BJ	BJ	BJ	A	AJ	AJ
Blowers									
Utility sets									
floor-mounted	C	CJ	CJ	C	C	CJ	C	C	C
Suspended centrifugal type	CH	CH	CHJ	CH	CH	CHJ	CH	CH	CH
Fan heads									
floor-mounted	A	AJ	AJ	A	AJ	AJ	A	A	AJ
Suspended tubular centrifugal and axial fans									
Suspended up to 25 hp	F	F	FJ	F	F	FJ	F	F	F
30 hp and larger	F	FJ	FI	F	FJ	FI	F	F	FJ
Floor-mounted-motor on/in fan casing	A	AJ	AJ	A	AJ	AJ	A	A	AJ
Floor-mounted arrangement, 1 or any separately mounted motor	A	AJ	AJ	A	AJ	AJ	A	A	AJ
Cooling towers and condensing units	AJ	AJ	BI	AJ	AJ	BJ	AJ	AJ	AJ
Isolated pipe 6-in diameter and smaller	F	F	FJ	F	F	FJ	F	F	F
Isolated pipe 8-in diameter and larger	FJ	FJ	FI	FJ	FJ	FJ	F	F	FJ
Risers 6-in diameter and larger in excess of 160 ft	EJ	EJ	EJ	EJ	EJ	EJ	Always Essential		

There is a good possibility that local codes that still accept static analysis without requirements for resilient snubber interfaces will be updated during the next few years as this would be a step toward the present state of the art. Therefore, all the restraints in the selection guide incorporate either minimum ¼- or preferred ¾-in-thick neoprene materials as cushioning between steel components. Actual observation of hard-stop statically designed systems has shown a high failure rate during minor to moderate earthquakes, such as Santa Barbara and Calaxio, where Richter magnitudes were only in the order of 4.

Dynamic analysis is recommended for snubber systems restraining equipment located at an elevation above 26 ft in zone 4 if it must remain operative during and after a quake and some equipment that can sustain damage repairable after 48 h. Dynamic analysis is also suggested above 26 ft in zone 3 in a few selected cases. Static analysis suffices in zones 1 and 2. The g forces at "111 ft and taller" suggested in the charts under type I, dynamic analysis, were developed from the UBC-79 formula appearing in section 2312g, formula 12-8 with values stepped back 0.7 and 0.4 for the lower elevations as based on recorded events.

$$F_p = ZIC_pW_p$$

where F_p = seismic force applied at center of gravity
Z = seismic zone coefficient
I = occupance importance factor
C_p = horizontal force factor
W_p = weight of equipment

Since $G_n = F_p$
$G_h = ZIC_p$

Z Factor UBC-79

Zone	1	2	3	4
Z	³⁄₁₆	⅜	¾	1

Factor I. Of the three listed values for I in Table 23K (UBC-79) in Appendix D, the number 1.25 was used as a conservative value for "Other Than Selected Facilities." The value is increased to 1.5 for selected facilities.

Factor C_p. Of the two listed values for C_p in Table 23J (OBC-79) in Appendix D, the number 0.8 was used as the more conservative value (see footnote d, Table 23J-UBC-79).

G_v. G_v is taken as 0.67 of G_v in accordance with standard practice.

The following grouping of nonproprietory seismic snubbing devices are readily available from various recognized national manufacturers. Test data or calculations should be made available showing that both the steel and neoprene materials are operating within safe stress limits.

Tubular snubbers are recommended rather than cantilevered Z-shaped angle brackets because it is much easier to control clearances, pad designs, and anchoring configurations and work with a standard range of certified products. General product descriptions are not to be construed as specifications.

Seismic snubbers.

TYPE A: All directional with replaceable molded neoprene washer bushings not less than ¼ in thick.

TYPE B: All directional with replaceable molded neoprene bushings not less than ¾ in thick.

Mountings.

TYPE C: Housed spring mounts incorporating snubbers as described in type A and with adequate clearances to allow for freestanding spring action except during seismic disturbances.

TYPE D: Neoprene mountings with built-in positive restraints in all directions suitable for bolting to concrete or steel.

TYPE E: A pipe riser or pipe guide anchor similar to D but designed specifically for pipe riser anchoring or restraint. Normally installed on structural steel supports and welded in position.

Steel cables and isolation hangers.

TYPE F: A minimum of four steel cables installed at 45° angles to all three axes. Cable attachments with standard fittings to avoid bends across sharp edges. Cables shall be used in conjunction with isolation hangers incorporating minimum ¼-in-thick neoprene seismic stops. Cables are normally located at the four corners of equipment or two each per pipe hanger location with directions alternating on successive locations.

Flexible connections.

TYPE G: Molded nylon reinforced double-arch or elbow connectors capable of accommodating all calculated movements. Braided stainless steel may be substituted for neoprene where temperatures, fluids, or pressures exceed neoprene ratings.

Steel platforms.

TYPE H: Suspended structural steel platforms braced to withstand seismic loads imposed by platform-mounted equipment.

Analysis.

TYPE I: *Dynamic analysis.* A dynamic analysis resulting in a maximum force level to the equipment of 4g. The model analysis should be based on a lumped mass system. Regional floor-by-floor spectrums should be part of the contract documents. In the absence of same, an input curve based on a uniform G response at all frequencies is suggested as per the G_h and G_v recommendations in the following tables.

Commercial Buildings

Elevation	Up to 25 ft				26–110 ft				111 ft and taller			
Zone	1	2	3	4	1	2	3	4	1	2	3	4
G_h	0.08	0.15	0.30	0.40	0.13	0.27	0.53	0.70	0.19	0.38	0.75	1
G_v	0.05	0.10	0.20	0.27	0.09	0.18	0.36	0.47	0.13	0.25	0.50	0.67

Essential Facilities Such as Hospitals, Fire and Police, Municipal Disaster Operation and Communication Centers Deemed to Be Vital in Emergencies

Elevation	Up to 25 ft				26–110 ft				111 ft and taller			
Zone	1	2	3	4	1	2	3	4	1	2	3	4
G_h	0.09	0.18	0.36	0.48	0.16	0.32	0.63	0.84	0.23	0.45	0.90	1.20
G_v	0.06	0.12	0.24	0.32	0.11	0.21	0.42	0.56	0.15	0.30	0.67	0.80

Note: In the State of California, Title 24 Table T 17 prevails suggesting a ground-level input of 1g with a 2g input in all upper floor locations.

TYPE J: *Static analysis* is performed using G_h and G_v from the charts in type I above or the State of California maximum requirements. Forces should be considered acting simultaneously at the center of gravity (CG) of the equipment to find the poorest direction including diagonals. Snubbers shall have minimum capability equal to the maximum combined loadings.

SEISMIC RESTRAINING DEVICES

There are two types of seismic restraints: active and passive. An active seismic device consists of one or more sensing elements that will trigger a set of lockout mechanisms to hold the equipment rigidly to the floor in the event of an earthquake. The sensing elements can be electronic or mechanical, and the lockout mechanism can be electrically, pneumatically, or mechanically actuated (see Figure 7.10). However, such devices not only involve high initial cost but also require regular maintenance such as periodic testing to see if the sensing elements and lockouts remain in good repair. It is difficult to enforce these checking procedures, which guard against earthquake damage, for the same reason that people tend not to fix their roofs when it is not raining. Since there is no obvious gain in keeping these systems operative, they are not as dependable as the passive systems that require no maintenance.

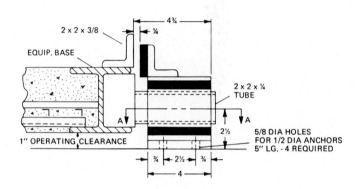
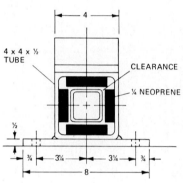
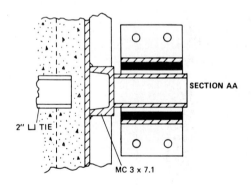

SEISMIC SNUBBER CAPACITY 2 KIPS

NOTE:

1. ALL NEOPRENE PADS TO BE **BRIDGE BEARING NEOPRENE** WITH A MINIMUM THICKNESS OF ¼". PADS TO BE CAPABLE OF A MINIMUM 1.0 G WITHOUT PERMANENT DEFORMATION.
2. CLEARANCE TO BE PRE-SET AT MANUFACTURER'S FACTORY WITH REMOVABLE SHIMS.
3. SEISMIC SNUBBERS TO LOCATED AND SET AFTER EQUIPMENT IS INSTALLED AND OPERATING TO ASSURE DESIGN CLEARANCES ARE MAINTAINED.
4. ALL DIMENSIONS ARE IN INCHES.

FIG. 7.7 Seismic restraint. *(Mason Industries.)*

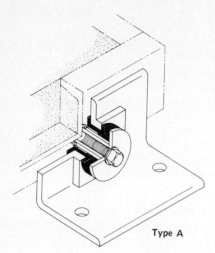

Type A

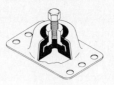

Type D

Type E

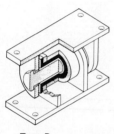

Type B

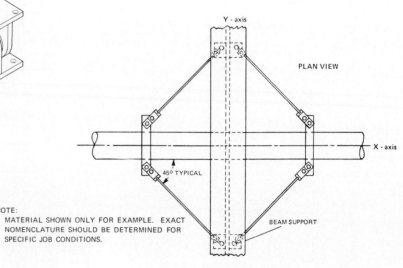

PLAN VIEW

Y - axis

X - axis

45° TYPICAL

BEAM SUPPORT

NOTE:

MATERIAL SHOWN ONLY FOR EXAMPLE. EXACT
NOMENCLATURE SHOULD BE DETERMINED FOR
SPECIFIC JOB CONDITIONS.

Type F

Type C

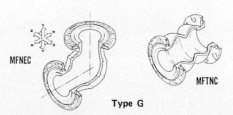

MFNEC

MFTNC

Type G

Passive seismic restraining devices are most commonly snubbers (see Figures 7.7 and 7.8). A snubber is simply a set of resilient pads or bushings and a pair of steel housings arranged so that when equipment travels beyond an established gap, the resilient pads will be engaged. The protected equipment is thus restrained and movements restricted to where they will not cause damage to the connected services while limiting the deceleration forces to the fragility levels.

FIG. 7.8 **Various types of seismic restraints.** *(Mason Industries.)*

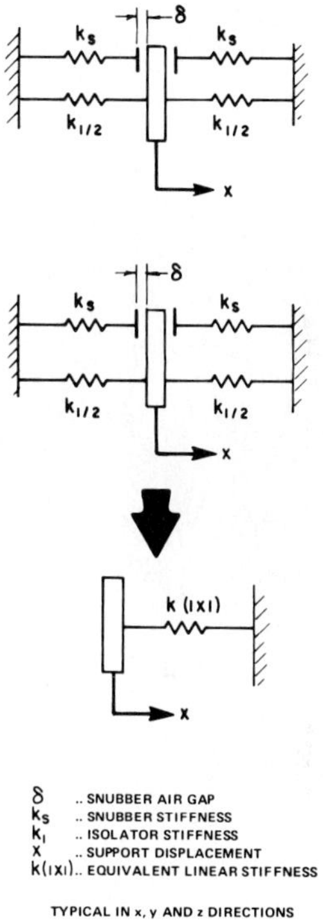

δ .. SNUBBER AIR GAP
k_S .. SNUBBER STIFFNESS
k_I .. ISOLATOR STIFFNESS
x .. SUPPORT DISPLACEMENT
$k(|x|)$.. EQUIVALENT LINEAR STIFFNESS

TYPICAL IN x, y AND z DIRECTIONS

Type I

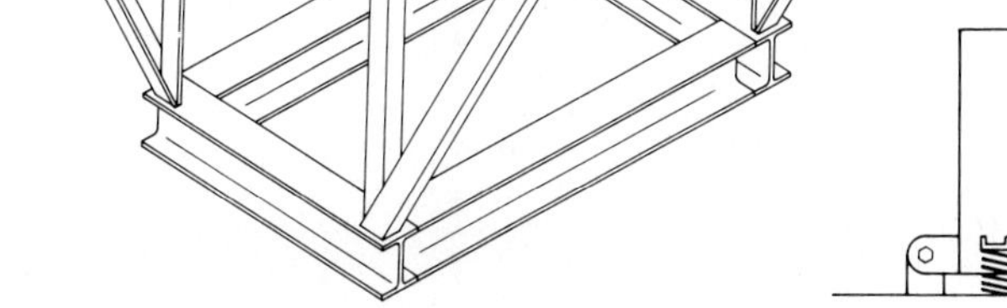

Type H

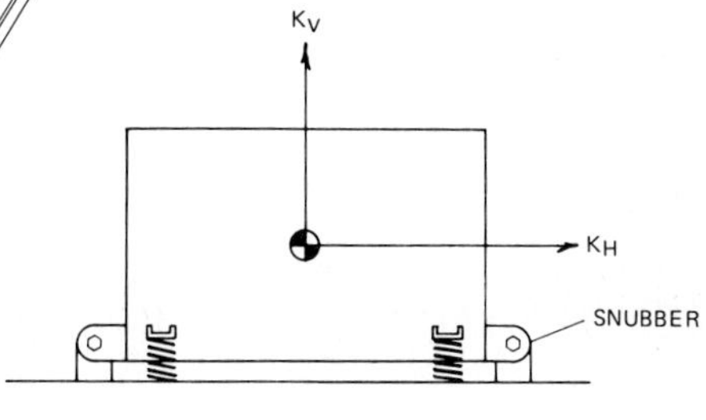

Type J

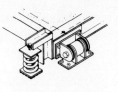

SNUBBER INSTALLED
ALONGSIDE EQUIPMENT

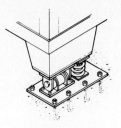

SNUBBER INSTALLED
UNDER EQUIPMENT

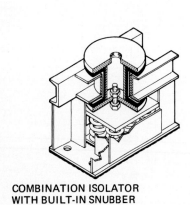

COMBINATION ISOLATOR
WITH BUILT-IN SNUBBER

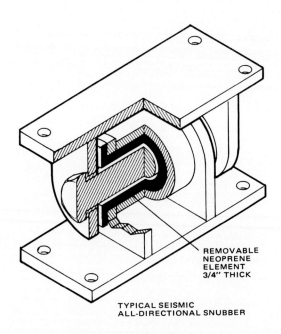

REMOVABLE
NEOPRENE
ELEMENT
3/4" THICK

TYPICAL SEISMIC
ALL-DIRECTIONAL SNUBBER

FIG. 7.8   (Cont.)

SEISMIC ISOLATION FOR RESILIENTLY MOUNTED EQUIPMENT

Once the decision has been made to mount equipment resiliently, an inspection should be made to evaluate the entire load path of the seismic input from the structure to the equipment. Included in such an evaluation should be (1) the attachment of the seismic restraint to the structure,* (2) the seismic snubbing ability to withstand the input,† (3) the attachment of the seismic restraint to the equipment or base frame, and finally (4) the equipment to the base itself.‡ Failure in any of these modes could be catastropic.

One factor to be weighed in selecting a seismic snubber is the equipment fragility level, or the maximum g level, that a piece of equipment can withstand before failing internally. Equipment such as pumps, fans, and air-handling units have fairly high fragility levels, often as much as 4 to 5g's. Electrical equipment such as transformers, dimmers, and switch gears may have tolerances to shock that are extremely low — ¼g to ½g. In between, equipment such as cooling towers, condensing units, and packaged equipment can withstand about 3g's.

Past tests and measurement by various equipment manufacturers have tended to establish these approximate levels of fragility.§ As may be noted from the g values described above, the electrical portion of a mechanical system is generally the weak link in seismic situations. Loss of some electrical equipment, such as the switch gear, renders the entire mechanical and electrical system inoperable. Therefore, the engineer should check with the equipment manufacturer to determine if a different type or model that performs the same function may be substituted, thereby increasing the system shock tolerance level.

---

* See Table 26G UCB 80 reproduced as Figure 7.12.
† *Guidelines for Seismic Restraints of Mechanical Systems*, SMACNA (see Appendix H).
‡ *AISC Steel Handbook*, Chapter 4. See Appendix B for tables of rivets and threaded fasteners.
§ See Smith-Emery test (Appendix C) and *Seismic Testing*, Wyle Laboratories.

7.10 This approach uses conventional vibration isolation with a lockout system that, at 0.03g acceleration, automatically triggers a series of devices which fasten the isolated equipment rigidly to the structure (Figures 7.9 and 7.10). By eliminating relative motion between the equipment and the structure, the forces felt by the equipment will be no more than those imposed on the building.

By automatically locking the resiliently supported system to the structure during an earthquake, damage can be greatly reduced and the chance that equipment will continue to function is greatly increased.

This lockout system consists of a preset 110/12 V electrical triggering mechanism, two 5-A 110-V external relays, one power interruption delay circuit, a 40-psi low-pressure system, a 12 V/100-psi solenoid valve, pneumatic tubing, electrical wiring, and the required number of air-activated mechanical lockout devices.

The system can be expanded to include one or more air reservoir bottles or tank-type air compressors and a multiple number of solenoid valves and/or lockout devices.

Operation. When the acceleration of the building structure reaches a preset level of 0.03g, the seismic trigger closes a line switch, which, in turn, opens the air solenoid and causes the locking pistons to advance into the centering sockets attached to the supported equipment. The line switch also activates a relay with external switching circuits so that sirens, lights, or other alarms may be activated.

The line switch remains closed until 10 s after the acceleration level drops below 0.03g.

In the event of a power failure, or low-air-pressure condition, after 15 s a separate NC relay closes, activating the air solenoid, closing the external switching relay, and causing the locking pistons to engage. After the locking pistons engage in the centering sockets, the pistons are held in locked position by a spring-loaded pawl, so that a failure of air pressure or electrical power would not unlock the device or reduce its effectiveness.

After activating, the electrical trigger will reset in 10 s. The locking pistons must be partially dismantled to be reset.

For maintenance purposes, the electrical connections to the solenoid can be disconnected, thus preventing inadvertant activation.

Note, too, that earthquake shock levels across properly designed snubbers will be amplified. In many cases, a 1g input will result in a 2½g output across a snubbing device, because of inertia effects of the building structure

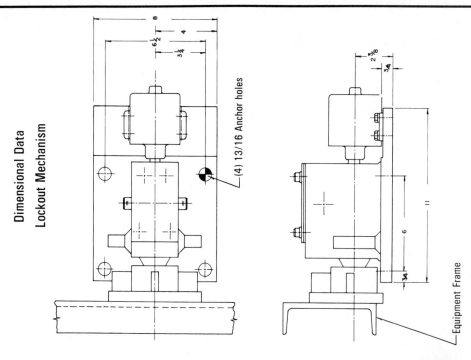

Dimensional Data
Lockout Mechanism

(4) 13/16 Anchor holes

Equipment Frame

FIG. 7.9 Seismic restraint lockout system. *(Peabody Noise Control.)*

on the equipment and because of the minimal air gap that must be introduced.

The snubber design itself raises other questions. Since an earthquake can approach a structure from any direction and can strike with the addition of vertical force components, the snubbing system employed must act approximately the same way in any direction. Uniformity of snubber response is also required because earthquake motion may include rocking as well as translational modes, and in some instances, the modes will be additive (in phase).

Final considerations are the cost of the seismic snubber and its established reliability. The optimum thickness of resilient material required to produce a reasonably reliable seismic snubber is ¾ in, and the material should be of at least bridge-bearing neoprene quality certified to AASHO specifications. In certain cases, one should have the material tested for authenticity, to conform with the following:

Neoprene Physical Properties			
Grade (durometer A)	40	50	60
Original physical properties			
Hardness ASTM D-676	40±5	50±5	60±5
Tensile strength, minimum psi ASTM D-412	2,000	2,500	2,500
Elongation at break, minimum percent	450	400	350
Accelerated tests to determine long-term aging characteristics			
Oven aging — 70 hrs., 212°F, ASTM D-573			
Hardness, points change, maximum	+15	+15	+15
Tensile strength, percent change, maximum	±15	±15	±15
Elongation at break, percent change, maximum	-40	-40	-40
Ozone (1 ppm in air by volume - 20% strain — 100 + 2°F — ASTM D-1149)			
100 hrs.		No Cracks	
Compression set			
(22 hrs./158°F, ASTM D-395 — method B)			
Percent, maximum	30	25	25

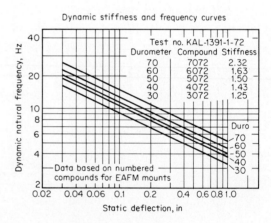

Dynamic stiffness and frequency curves

Test no. KAL-1391-1-72

Durometer	Compound	Stiffness
70	7072	2.32
60	6072	1.63
50	5072	1.50
40	4072	1.43
30	3072	1.25

Data based on numbered compounds for EAFM mounts

Once a building and its mechanical systems have been subjected to a seismic event, a quick inspection of snubbers should be performed, since one is never sure if the seismic event is a foreshock or if there will be numerous aftershocks.

Many people tend to associate the term "seismic" with massive earthquakes. To avoid this error, remember that a difference on the Richter scale between 7 and 8 means an increase in energy release of approximately 30 times.

Formula: $\log E_s$ (ergs) $= 11.8 + 1.5m$ (Richter)

Thus, in designing and installing earthquake protection for resiliently mounted equipment, the purpose is not to

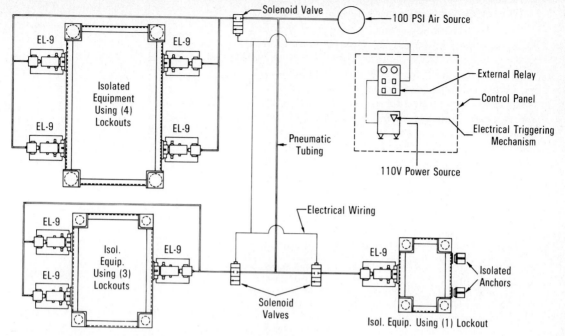

FIG. 7.10 **Seismic restraint lockout diagram.** *(Peabody Noise Control.)*

protect for major or massive earthquakes within 20 mi of
the epicenter (7 or higher on the Richter scale). Rather, it
is an attempt to keep equipment close enough to its
original position so as not to have damaged or broken
service connections and to keep it operational once it has
received a seismic shock.

Both professional engineers and laypersons raise the
question of how much vibration protection the equipment within a structure is required to have. With a few
important exceptions, this question was answered
several years ago by Mason Industries through its VCS
lecture series and accompanying isolation selection chart
and later in Bulletin SCS-1000.* It is important to use
only a professional engineer trained in earthquake
engineering to set up the guidelines for seismic projections. Unlike vibration control, where mistakes merely
cause inconvenience and the possible loss of time,
incorrect seismic protection can cause the loss of life and
property. For this reason, this manual includes a basic
guide table for engineers (not laypersons) to use in
selecting proper seismic protection for resiliently
mounted mechanical equipment (see Figures 7.3 to 7.6).

_____
* Mason Industries Bulletin SCS-100.

In all cases, this guide exceeds the existing building codes, which are only minimal guides and are therefore potentially misleading and possibly inadequate in some instances.* The guide table attempts to provide a reasonable solution, both technically and economically, to the complex problem of keeping a building operational shortly after a minor-to-moderate earthquake. The author makes no claim that the use of this guide will prevent damage as a result of a major seismic event. Should an earthquake occur in the Richter 7 or greater range, the best one can expect is to prevent equipment from falling off its mountings and causing severe structural damage.

---

* Uniform Building Code, U.S., Earthquake Regulations, Section 2312. National Building Code, U.S., Earthquake Loads, Section 907. National Building Code of Canada, Effects of Earthquakes, Subsection 4.1.7. Appended excerpts from the UBC for seismic zones appear in the Appendix D. Also excerpts from Building Standards—State of California for Hospitals appear in Appendix F and for schools in Appendix G. Appendix E contains pertinent excerpts from Register 74, Public Health—Safety of Construction of Hospitals.

7.11 Other Design Aids

7.12 To aid the designer of seismic restraint systems in determining what earthquake zone the project is in, two maps by two different authorities are provided in Figure 7.11, and a table of allowable shear and tension of bolts is given in Figure 7.12.

FIG. 7.12 **Allowable shear and tension of bolts.**

Diameter, in	Minimun Embedment, in†	Shear* Minimum Concrete Strength, psi 2000	Shear* Minimum Concrete Strength, psi 3000	Tension* Minimum Concrete Strength, psi 2000 to 5000
1/4	2½	250	250	200
3/8	3	550	550	500
1/2	4	1000	1000	950
5/8	4	1375	1500	1500
3/4	5	1470	1780	2250
7/8	6	1790	2075	3200
1	7	1790	2075	3200
1⅛	8	1790	2250	3200
1¼	9	1790	2650	3200

* Values shown are for work without special inspection. Where special inspection is provided values may be increased 100 percent.

Values are for natural stone aggregate concrete and bolts of at least A307 quality. Bolts shall have a standard bolt head or an equal deformity in the embedded portion.

Values are based upon a bolt spacing of 12 diameters with a minimum edge distance of 6 diameters. Such spacing and edge distance may be reduced 50 percent with an equal reduction in value. Use linear interpolation for intermediate spacings and edge margins.

† An additional 2 in of embedment shall be provided for anchor bolts located in the top of columns for buildings located in seismic zones 2, 3, and 4.

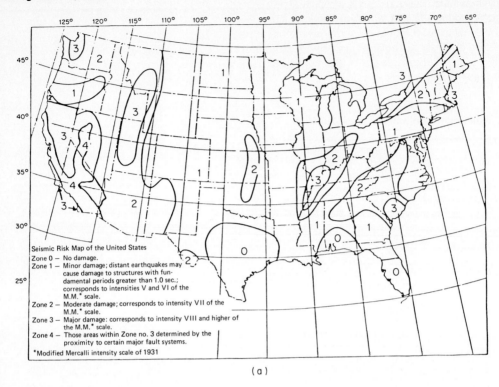

(a)

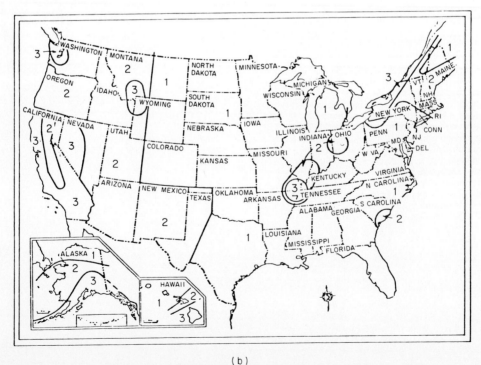

(b)

FIG. 7.11   Earthquake zone maps: *(a)* uniform building code and *(b)* national building code.

SEISMIC ISOLATION AND PROTECTION OF EXTERNAL ELEMENTS

So far, we have discussed seismic protection of mechanical and electrical equipment. Such equipment has some external connections that tie it into the entire building system. A discussion of these external elements—electrical conduits, waste pipes, sanitary lines, and polyvinyl-chloride (PVC) or other flexible materials—is included here. Many areas still require further engineering studies and research. While snubbing of equipment is basically solved in terms of the available hardware and its application, this is not the case for the external elements discussed here. The following figures and text show examples of installations of seismic protection for mechanical and electrical equipment and their external connections.

7.13 Isolator for Gas-Engine-Drive Chiller

A minimum 8-in-thick concrete inertia block had been specified for this installation. However, the equipment manufacturer, concerned about the critical alignment of the gear reducer and the engine, requested that a dynamic analysis be performed to ensure that the base construction would be of the required thickness. The analysis showed that a concrete inertia block, suitably reinforced and 2 ft thick, was required. Thus, in this installation, the inertia block weighs about 100,000 lb, while the equipment is approximately 60,000 lb.

Spring isolators were recessed into pockets in the inertia block because of floor space limitations and the actual configuration of the base. Open, stable, steel springs are employed here, with 3-in static deflection capability. The seismic snubbers are all directional type with ¾-in neoprene bushing materials.

Piping connections to the chiller were fitted with flexible pipe fittings. One such fitting may be seen in the smaller pipe at the lower right corner of Figure 7.13 (see arrow). Also visible are the flexible sections of electrical conduit between the control panel and chiller components.

FIG. 7.13   Isolator of gas-engine-drive chiller.

7.15 Seismic snubbers installed for the pump vibration-isolation system in Figure 7.14 are adequate. They are placed at four opposing points around the pump base, thus effectively resisting rotational reactions from the pump and its base during a seismic event. See Figure 3.38*b* for recommended location of the seismic restraints.

The spring vibration isolators do not need to be bolted down in installations of this type. The direct bolting through the base plate of the isolator (shown in Figure 7.14), without resiliently separating the bolt from the isolator, makes the neoprene pad that is secured to the underside of the isolator ineffective.

When bolting is required, Table 26G UBC reproduced here as Figure 7.12, should be consulted for proper bolt selection.

Note that the pump inlet and discharge pipe elbows are supported rigidly and directly from the pump inertia block, as is commonly recommended. In such applications, the piping connected to the pump is usually hung or mounted on resilient vibration isolators for a given distance from the pump. The neoprene pads that show under the pipe elbow support plates are not necessary.

Should a significant earthquake occur, causing the pump shown in Figure 7.15 to move downward, the white PVC pipe running under the pump base could be crushed. As it is installed here, the pipe somewhat hinders the vibration-isolation system from operating at top efficiency.

FIG. 7.15   **Plastic pipe under pump base.**

FIG. 7.14   **Vibration-isolated seismically restrained.**

Along with the air springs, seismic snubbers are shown
in Figures 7.16 and 7.17. Note that the attachment of the
snubber to the equipment base is near the imaginary
horizontal centerline of the I-beam structural shape. In
such installations, during a seismic event, the snubber
will tend to twist the I beam because of torsional loads
imposed by the earthquake shock. To prevent this, a
backing beam or other adequately sized structural shape
should be installed to resist the torsional force. A
practical alternative is to install seismic snubbers at the
corners of the equipment base, where the equipment
base frame members already provide the strength to resist torsional forces.

In the foreground of Figure 7.17 is a rotary valve (see
arrow). These valves, used for leveling the air springs,
are provided with a built-in time delay. However, these
valves must be sensitive enough to react before the
equipment is allowed to move ¼ in.

Figure 7.17 shows the rotary valve linkage. Where the
vertical linkage strap connects to the isolator base plate a
neoprene vibration-isolating pad should be provided. In
Figure 7.17, the isolator base plate bolts have diameters of
about ⅜ in, with an allowable shear of 550 lb per bolt.
This bolting is adequate because the seismic snubber is
rated for 350 lb. The clearance under the equipment
base, however, is not adequate. A 1-in minimum is
usually satisfactory, with ¼-in clearance at the extended
corner plates.

Small copper (or PVC) air lines required for air springs
should be kept inboard, close to the equipment base, for
maximum protection and for preventing damage to the
air lines.

FIG. 7.16   Air mount with seismic snubber.

FIG. 7.17   Air mount with seismic restraint.

7.18 Correcting a Vibrating Transformer

7.19
7.20 Figures 7.19 and 7.20 show the results of remedial work applied to a noisy, vibrating transformer with a stamped metal enclosure. The original installation for this transformer/cabinet package included only rubber pads under the unit. So much noise and vibration traveled to noise-sensitive occupied spaces below the transformer that authorization was given to correct the annoying situation.

Examination showed that it was impractical to raise the whole equipment package more than ½ in. Thus the core was cut away from the cabinet and mounted on air springs. The cabinet, which was then subject only to air excitation, could be isolated on neoprene mounts. The supplemental structural steel shape shown behind the stamped metal enclosure was added to provide additional support for the transformer core within the casing. (See arrow 1, Figure 7.19.)

FIG. 7.18 **Transformer snubber (nonseismic).**

Figures 7.19 and 7.20 show the air springs with built-in seismic restraints. The horizontal steel tube section running over the top of the air spring nestles into the vertical steel uprights on each side of the isolator. The vertical uprights have neoprene lining inside, providing a resilient separator between the uprights and the horizontal steel tube snubber. Full structural welds are used throughout to give sufficient strength.

Although it is not clearly visible in Figure 7.19, spilled lubricating oil had damaged the air bag. To prevent this problem, do not locate rotary valves (see arrow 2) directly above the air bag. An acceptable alternative, if the rotary valve must be placed above the air bag, is to provide a protective shroud over the bag itself.

These combination units were secured to the building by drilling through the mechanical equipment floor slab and extending the hold-down bolts all the way through the slab to steel spreader plates on the underside. This method of installation gave the bolting the required seismic resistance capability.

An added safety feature has been applied here in the form of a button valve (see arrow 3), which can be seen in Figure 7.19 under the steel bar stock bracket at the top of the air spring. Should the bus valve and the snubber fail and the air spring rise beyond ⅛ in, the button valve is touched, releasing air to the atmosphere and preventing any further upward movement.

The seismic snubbing device is shown in the lower right corner of Figure 7.20. The neoprene mount just to the left of the seismic unit is for vibration-isolating the transformer enclosure only. The seismic unit controls both horizontal and vertical movement.

Figure 7.18 is a version of a nonseismic snubber.

FIG. 7.19　Air mount with built-in seismic restraints.

FIG. 7.20　Air mount with snubber and separate seismic restraint.

Neoprene pads attached to the back of the angle brackets provide the resilient separation required between the bracket and the equipment base (see arrow). This type of snubber provides protection for horizontal movement only.

7.21 Seismic Restraints for Electrical Transformer

The resilient seismic restraint unit in Figure 7.21 uses a section of steel tube with a short angle-iron clip welded to each side of the tube. These angle clips, when bolted to the building, serve to limit horizontal and vertical movement of the equipment during a seismic event.

The surface of the angle-iron legs facing the equipment base should be covered with a neoprene or rubber pad similar to that used to line the steel tube section. The other two angle-iron clips, welded to the bottom sides of the tube, are used to secure the unit in place by bolting into the concrete floor slab.

Neoprene or rubber pads are cemented on all four inside surfaces of the steel tube for its full length. A piece of steel bar stock is then centered in the resiliently treated tube and welded to the equipment base. The bar stock section is clearly seen protruding through the center of the tube. For detail, see Figure 7.7.

The waffle or ribbed neoprene pads that can be seen under the equipment to the left and to the right (see arrow) of the seismic restraint unit are the vibration isolators for this electrical transformer casing.

FIG. 7.21 **Seismic restraints for electrical transformer.**

7.22 Seismic Restraint Isolators
7.23 with Travel Limit Stops
7.24

Figures 7.22 and 7.23 show vibration isolators with built-in vertical travel limit stops. This type of unit is designed to be hard-bolted in place at its base plate. Therefore, the addition of neoprene pads under the base of the unit is not required in Figure 7.22. Where the bolts are too close to the edge of the isolator base to meet the requirements of prevailing codes steel plates may be anchored to the structure and the steel isolated base plate either bolted or welded to the larger structural plates. These isolators have internal springs sitting in neoprene cups top and bottom. See Figure 7.24 for a cutaway view.

In Figure 7.22, the vertical adjusting bolt can be seen touching the side of the top hole of the isolator housing (see arrow). It is important to center the unit carefully before bolting it in place. Isolators with oversized clearance holes are useful. Such isolators are now available from most manufacturers. The vertical clearance between the adjusting nut and isolator top housing should be set at about ¼ to ½ in for preventing excessive vertical movement of the isolated equipment.

In Figure 7.23, note that hold-down bolts are placed too far out on the base plate (see arrows). Similarly,

FIG. 7.22 Captive-type isolators.

FIG. 7.23 Isolator bolted in place.

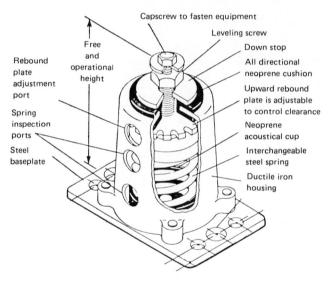

FIG. 7.24 Cutaway view of captive isolator. *(Mason Industries.)*

hold-down bolts should not be placed close to the edges of the concrete housekeeping pads. Recommended dimensions are given in the *Uniform Building Code.** The necessary flexible electrical connection to the equipment motor is also shown in Figure 7.22.

————————————

* See footnote * in Figure 7.12.

7.25   Isolation for Air Compressor

7.26   Seismic mounts are employed in Figure 7.25 to provide
vibration isolation under this tank-mounted air compressor. The mounts are set on a concrete housekeeping pad,
and instead of using a concrete inertia block, the isolators are placed directly under the four feet of the
equipment package. Then each isolator is bolted to the
building. These vibration isolators are similar to that
shown in Figure 7.26.

FIG. 7.26   **Shipboard-type resilient mount.** *(Kine*
matics Division of Lord Corp.)

FIG. 7.25   **Isolation for air compressor.**

For seismic snubbing devices to function satisfactorily
during an earthquake, the attachments to the building
structure must be designated and installed properly. In
Figure 7.27, note that the vertical limit rods at the sides of
the air spring are too small in diameter to sustain the
anticipated seismic shock.

The number of seismic restraints in this installation is
inadequate, and this arrangement provides only for
horizontal forces. During a seismic event, the mounted
equipment would tend to rotate around the axis provided
by the snubbers (see rotational arrows). Additional
snubbers, installed perpendicular to the existing axis,
will resist and limit such rotational movement.

The structural steel members used in these snubbers
are inadequate and do not conform with recommendations published in the *Uniform Building Code*. The
angle-iron leg for the floor attachment with the slot for
the hold-down bolt is not recommended. These slots
allow slippage during seismic shocks; thus a hole, not a
slot, is recommended.

Neoprene pads used to provide resilient separations
between adjacent steel members of seismic snubbers
should be a minimum of ⅝ in thick. Note carefully that
the steel bracket for the air springs has a vertically
extended leg to provide the recommended ¼-in clearance
between the bracket and the floor or, in this case, the
housekeeping pad below (see arrow).

FIG. 7.27 **Seismic snubber and air mounts.**

7.28 Seismic Restraint for Air-Handling
7.29 Units in Nuclear Plants _____

7.30 The equipment in Figures 7.28 to 7.30 is special. Notice the massive size of the seismic snubber in contrast to the open steel spring vibration isolator in Figure 7.29. This floor-mounted air-handling unit in a nuclear power generating plant is being seismically protected for 6g's, or 6 times the weight on the spring. In this case, the weight is 1350 lb, so 6g's = 8100 lb. This degree of protection will satisfy the seismic category I at this plant location. The seismic snubber in Figure 7.28 is welded to the large steel base plate, which is properly secured to the concrete building structure.

The air-handling equipment being protected in Figures 7.28 and 7.30 is suspended from overhead on substantial steel framing, and vibration isolation is provided by neoprene mounted on horizontal steel supporting shapes (see arrow in Figure 7.28). This type of seismic protection is about 20 times more costly than conventional steel cables and clamps. However, such protection is necessary for this very special installation.

FIG. 7.28 Seismic restraint for air-handling unit in a nuclear plant.

FIG. 7.30 Seismic restraint and neoprene isolator for air-handling unit hung from above.

FIG. 7.29 Seismic restraint unit and spring isolator for floor-mounted air-handling unit.

7.31 Turnbuckles and Cable Restraints
7.32

Turnbuckles in steel cable seismic restraints are not recommended. They represent a weak link in an otherwise sufficiently strong system. See arrows for turnbuckles in the seismic restraints for the pipeline in Figure 7.31.

The seismic cable restraint shown in Figure 7.32 illustrates an extremely hazardous condition. The slack steel cable passes just below and very close to a bank of bus bars carrying several high-voltage electrical conductors. In a seismic event, the cable can snap taut, thoroughly rupturing the bus bars (see arrow), possibly contributing to a serious power outage, and even causing the piping to become electrified. Generally, such steel cable restraints should be at least 3 in away from other utilities or systems or parts of the building structure.

Wire rope size and strength requirements for seismic restraint capabilities may be determined and selected from the *Machinery's Handbook*, 20th ed., Industrial Press Incorp., New York, 1978, pages 485 to 501. These pages are reprinted by permission in Appendix A.

Tables for rivets and fasteners are reprinted by permission in Appendix B.

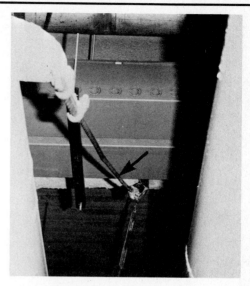

FIG. 7.32 **Piping cable restraint.**

FIG. 7.31 **Piping partially restrained.**

7.35 The preferred method for providing seismic restraints on
7.36 hung piping systems is clearly seen in Figures 7.33 and
7.34. In this case, the restraint employs steel cable,
thimbles at attachment loops, and a cable clamp at the
main cable 360° loop (see arrow 1). See Figure 7.35 for
details of cable thimbles. (See Appendix A for wire rope
sizes and strengths.)

The full 360° loop in the cable restraint allows the
installer to provide the degree of slackness that will
permit the vibration isolators to perform acceptably and
limit the amount of pipe movement due to an earthquake
to an acceptable quantity. The cable loop remains intact
during a seismic event and is not intended to release.

Looking closely at Figure 7.33, you can observe the
second cable (see arrow 2) extending from the concrete
beam attachment back to the pipe attachment (not seen).
This method, coupled with two cables symmetrically
arranged on the other side of the pipe, is used to furnish
seismic protection in both the X and Y axes.

A typical cable and fitting selection table and a
drawing detail for the cable and thimble restraint are
shown on Figure 7.35. Figure 7.36 gives earthquake
restraint construction details for a steel cable connector.

FIG. 7.33 **Piping cable restraint.**

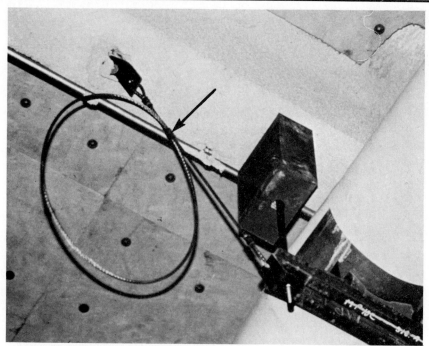

FIG. 7.34   Pipe cable restraint with hanger isolators.

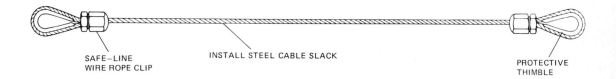

SAFE—LINE
WIRE ROPE CLIP

INSTALL STEEL CABLE SLACK

PROTECTIVE
THIMBLE

Typical Earthquake Restraint Construction Details for Steel Cable Connectors

Cable Diameter (in)	1/8	3/16	1/4	5/16	3/8	7/16	1/2
Maximum Load (lbs)	1000	2100	3500	4900	7200	10,200	13,300

FIG. 7.36   Earthquake restraint construction details for steel cable connector. *(Amber/Booth Company.)*

Wire rope clips

"First grip" wire rope clips

Wire rope clips are manufactured to rigid specifications. Forged base with high wings holds ropes snugly in place, one over the other. Rope channel, shaped to assure perfect non-slip union. Entire clip — hot dip galvanized to resist corrosive and rusting action.

The "Fist Grip" Clip's bolts are an integral part of the saddle, with nuts on opposite sides of the clip, enabling the operator to swing his wrench in a full arc for fast installation. The "Fist Grip" Clip is easily installed since its two parts are identical. Forged steel, hot dip galvanized with standard heavy hex nuts.

Clip and Rope Size Inches	Minimum Number Clips Required	Weight Pounds Per 100
1/8	2	5
3/16	2	9
1/4	2	18
5/16	2	30
3/8	2	42
7/16	2	70
1/2	3	75
9/16	3	100
5/8	3	100
3/4	4	150
7/8	4	240
1	5	250
1 1/8	6	310
1 1/4	7	460
1 3/8	7	520
1 1/2	8	590
1 5/8	8	730
1 3/4	8	980
2	8	1340
2 1/4	8	1570
2 1/2	9	1790
2 3/4	10	2200
3	10	3200
3 1/2	12	4000

Clip and Rope Size Inches	Minimum Number Clips Required	Weight Pounds Per 100
3/16–1/4	2	21
5/16	2	27.08
3/8	2	45
7/16	2	65
1/2	3	65
9/16	3	112.5
5/8	3	112.5
3/4	3	143.75
7/8	4	200.6
1	5	260
1 1/8	5	325
1 1/4	6	425
1 3/8	6	700
1 1/2	7	700

Thimbles

Extra heavy
A rugged rope thimble recommended for hea
duty service

Standard
Recommended for light duty service

Stainless steel

Cable sleeves

FIG. 7.35   Wire cable fittings. *(The Crosby Group.)*

7.37 Seismic Cable Restraints for Hung Piping

7.38
7.39
7.40
7.41
Figure 7.38 shows the practical use of cable thimbles (see arrow 1) (no turnbuckles) for attachments of cables to pipe and building structure. A word of caution is appropriate here. Care should be taken to drill holes for cable attachments inbound on the steel pipe clips (see arrow 2) as far as the thimble shape will allow without binding the thimble.

In Figure 7.37, assume that a centerline along the pipe represents the X axis and that the Y axis is perpendicular to it. As installed, the seismic cable restraints offer no capability on the Y axis. When designing and installing such seismic devices, one should allow for the cables to run out from the pipe attachments at between 30 and 60° angles, to achieve seismic control for both X and Y. See Figure 7.39 for an example. See Figure 7.40 for an alternate method of installing snubbers.

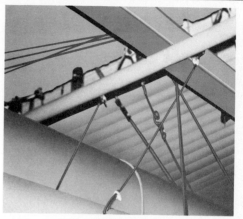

FIG. 7.37 Piping cable restraints.

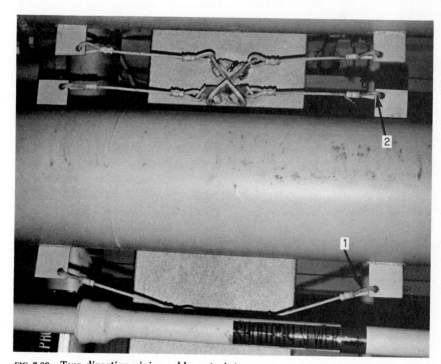

FIG. 7.38 Two-direction piping cable restraints.

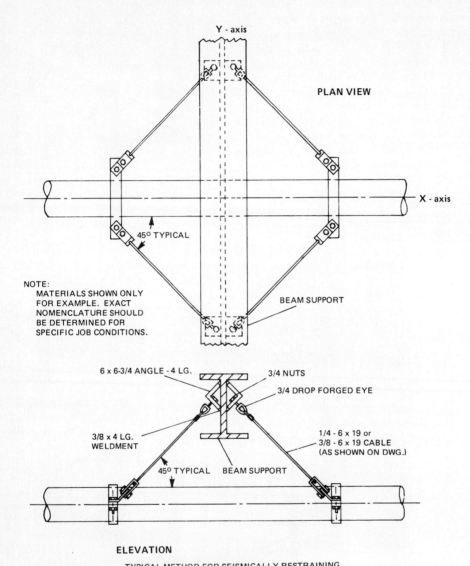

PLAN VIEW

Y - axis

X - axis

45° TYPICAL

NOTE:
MATERIALS SHOWN ONLY
FOR EXAMPLE. EXACT
NOMENCLATURE SHOULD
BE DETERMINED FOR
SPECIFIC JOB CONDITIONS.

BEAM SUPPORT

6 x 6-3/4 ANGLE - 4 LG.

3/4 NUTS

3/4 DROP FORGED EYE

3/8 x 4 LG.
WELDMENT

1/4 - 6 x 19 or
3/8 - 6 x 19 CABLE
(AS SHOWN ON DWG.)

45° TYPICAL   BEAM SUPPORT

ELEVATION

TYPICAL METHOD FOR SEISMICALLY RESTRAINING
SUSPENDED PIPING USING STEEL CABLES

FIG. 7.39   Typical method for seismically restraining suspended piping using steel cables.
(Mason Industries.)

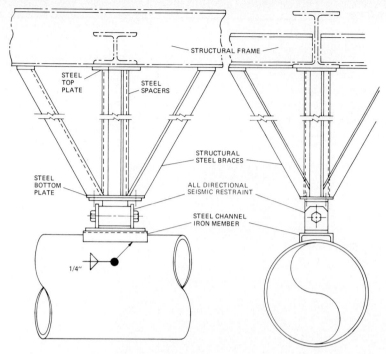

FIG. 7.40 **Alternate method for installing seismic snubbers on piping.** *(Mason Industries.)*

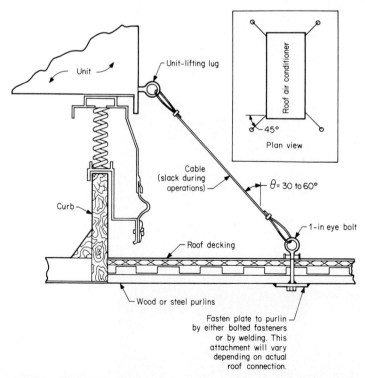

FIG. 7.41 **Suggested method for seismic restraints for curb-isolated rooftop equipment.** *(Mason Industries.)*

7.43 The cooling tower in Figure 7.42 is isolated on open steel
7.44 spring mounts with built-in travel limit stops. The
7.45 housing of the isolation unit is still engaged, as can be
seen by observing the right-hand top side support of the
isolator (see arrow). Until a clearance of about ¼ to ½ in
is established between the top of the side supports and
the top plate of the isolator unit, the mount cannot
provide the vibration isolation intended.

FIG. 7.42   **Spring with travel limit stops.**

FIG. 7.43   **Cable restraint connections to pipe.**

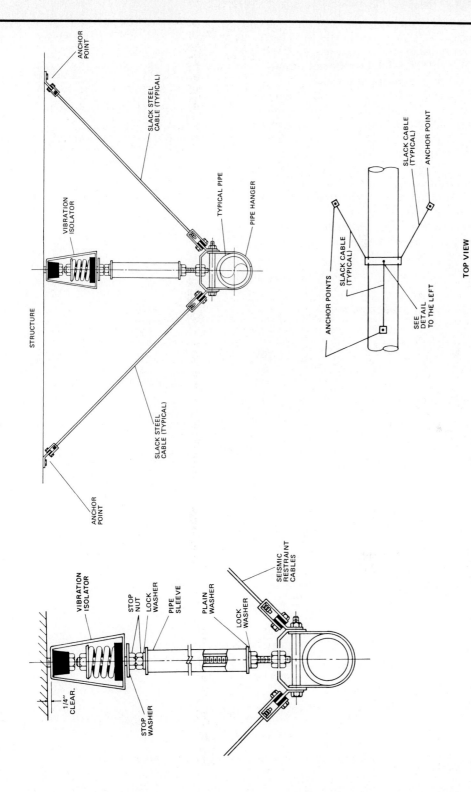

ANCHOR
POINT

SLACK STEEL
CABLE (TYPICAL)

VIBRATION
ISOLATOR

TYPICAL PIPE

PIPE HANGER

STRUCTURE

SLACK STEEL
CABLE (TYPICAL)

ANCHOR
POINT

SLACK CABLE
(TYPICAL)

ANCHOR POINT

ANCHOR POINTS

SLACK CABLE
(TYPICAL)

SEE
DETAIL
TO THE LEFT

TOP VIEW

VIBRATION
ISOLATOR

STOP
NUT

LOCK
WASHER

PIPE
SLEEVE

PLAIN
WASHER

LOCK
WASHER

SEISMIC
RESTRAINT
CABLES

1/4"
CLEAR.

STOP
WASHER

FIG. 7.44 Typical bracing for vibration-isolated pipe. (*Mason Industries.*)

Here again, cast or malleable turnbuckles have been employed in the steel cable seismic pipe restraints. Easily fractured devices such as turnbuckles are not recommended as part of a seismic restraint system.

Another installation detail that significantly reduces the protection capabilities of a steel rope seismic snubber is shown in Figure 7.43. Note that where the steel rope is connected through the steel bracket plate (see arrow) welded to the pipe that no protective thimble has been used. The rough and sharp edges of the bracket plate can now more easily cut through the strands of the wire rope when an earthquake shock occurs. The recommended policy is to always use wire rope thimbles for these connections. Alternative methods for seismic restraints and vibration isolation for both single and multiple pipes are shown in Figures 7.44 and 7.45

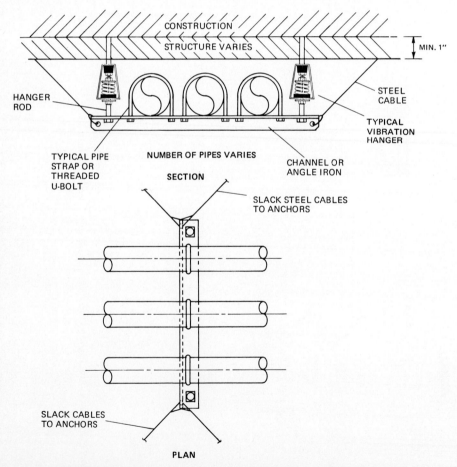

FIG. 7.45 **Lateral and longitudinal restrained piping on trapeze with vibration-isolation hangers.** *(Mason Industries.)*

7.46 Isolators for Vane Axial Fan

7.47 This fan has been floor-mounted on double-deflection neoprene vibration isolators. A floated floor that comes up to and around the concrete housekeeping pad for the fan has been provided. The structural break between the floor and the housekeeping pad (see arrow) caused by the resilient sealing compound installed in the joint at the floor surface can be detected by close examination.

Two seismic restraints are not an adequate number for this fan, because it leaves the fan free to rotate (see rotational arrows) around the Y axis during an earthquake. (See Figure 7.48 for close-up of snubbers.) A minimum of four snubbers is recommended for such installations, in order to handle the six degrees of freedom present. These are the three translation axes (x, y, and z) and the three rocking modes (yaw, pitch, and roll).

When fans similar to the type shown in Figure 7.46 are hung or suspended from above and earthquake protection is required, an acceptable method for seismic restraints is that shown on Figure 7.47.

FIG. 7.46   Isolators for vane axial fan.

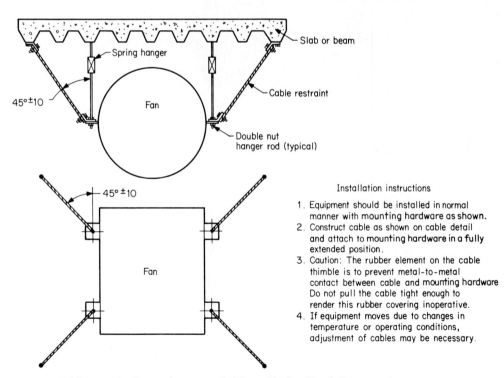

- Slab or beam
- Spring hanger
- Cable restraint
- Double nut hanger rod (typical)
- Fan
- 45°±10
- 45°±10

Installation instructions

1. Equipment should be installed in normal manner with mounting hardware as shown.
2. Construct cable as shown on cable detail and attach to mounting hardware in a fully extended position.
3. Caution: The rubber element on the cable thimble is to prevent metal-to-metal contact between cable and mounting hardware Do not pull the cable tight enough to render this rubber covering inoperative.
4. If equipment moves due to changes in temperature or operating conditions, adjustment of cables may be necessary.

FIG. 7.47   Cable restraint layout for suspended fans. *(Amber/Booth Company.)*

7.48 Seismic Snubber Attachment

The type of seismic snubber attachment shown in Figure 7.48 is not recommended. Bolts that fit into slots can be shaken or torn loose during a seismic event. The primary purpose of such earthquake protection devices is to keep the isolated equipment in place. The flat-plate snubber attachment to the equipment base is inadequate and does not conform to established engineering criteria. The angle-iron floor attachment itself is very lightweight and can easily be bent or deformed by a seismic shock.

FIG. 7.48   Seismic snubber attachment.

7.49 Vibration Isolators with Travel
7.50 Limit Stops

7.51 Experience with many isolation units installed in buildings has shown that redundancy of spring adjustment bolts beyond the more practical three-point or tripod arrangement is superfluous. Figure 7.49 shows one of many such instances, revealed during many job installation inspections. The item to note is that one of the four adjusting bolts has backed down (see arrow) by itself due to difficulty in keeping a four-point loading arrangement perfectly equalized between the four points. The three-point or tripod arrangement eliminates this problem and the need for an extra bolt. See Figure 7.50 for recommended installation and equipment. Note also that a clearance of about ⅜ in has been achieved between the side support travel limit stops and the isolator top plate.

The multiple layers of ribbed or waffle-pattern neoprene pads under the base plate of the isolation unit can be effectively used to dampen out some of the higher-frequency vibration that may travel down the spring coils. However, if the isolator is hard-bolted, as is the case here, the neoprene pads become short-circuited or bypassed. A suggested solution to this is to include the isolation pads directly beneath the spring base plate within the housing rather than underneath. With this arrangement, the housing can be hard-bolted or welded in position. Should the inclusion of pads within the housing become impractical, the acoustical treatment recommended for making such hold-down bolts resiliently protected was shown earlier in Figures 3.26a and 3.33.

In this application of multiple layers of neoprene pads, fairly lightweight steel shim plates were provided between individual layers of neoprene so as not to exceed the allowable deformation of the pad layers, which is 15 percent of the pad thickness. For an illustration of controlling the deformation of neoprene pads, see Figure 7.51.

FIG. 7.49   Four-point spring isolator.

FIG. 7.50   Three-point spring isolator.

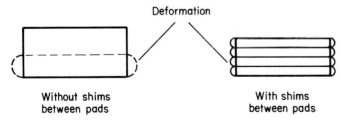

FIG. 7.51   **Multiple neoprene pads.**

7.52 Equipment Isolation with Horizontal Snubbers

Figure 7.52 illustrates base-mounted pumps on concrete inertia blocks that are vibrationally well isolated, i.e., open, stable, steel springs with cantilevered, side support brackets contributing to overall low center of gravity for the equipment and its base.

The snubbers just behind each of the spring isolators provide protection only in the horizontal plane. The bolting arrangement of the floor bolts is such that the eccentric load imposed by a seismic shock could well exceed the limitations of such a fastening. It is preferred to stagger the bolts or even place them on the centerline of the neoprene snubbing element.

FIG. 7.52 **Equipment isolation with horizontal snubbers.**

7.53 Field-Fabricated Snubber (Not a Seismic Device) _____

Figure 7.53 shows a field-fabricated snubbing device that is providing protection only in the horizontal plane. The single hold-down bolt inside the steel channel base can hardly be expected to hold fast during a seismic shock. Resiliency of the snubber is maintained by the ribbed neoprene pad between the vertical face of the snubber and the equipment base. (See Figure 3.38b for a better method of snubbing.)

FIG. 7.53   Field-fabricated snubber (not a seismic device).

7.54 Electric Transformer

The electric transformer shown in Figure 7.54 is vibration-isolated without seismic protection from the structure in this building, located in number 4 seismic risk zone (see Figure 7.11).

Seismic protection should be installed here, or if the transformer is not located in a noise-sensitive area, it could perhaps be hard-mounted—that is, bolted in place directly to its concrete housekeeping pad.

It is extremely dangerous to provide only vibration isolation without seismic protection for such equipment in earthquake zones with high seismic probability.

FIG. 7.54　Electric transformer.

7.55 Position Hanger

The isolators shown in Figure 7.55 are generally referred to as a position hanger. It is important to note that in this case the spring is noticeably overloaded and almost completely collapsed, so it was not possible to shift the load off the platform and on to the spring. If this hanger were of the proper capacity, the spring would not be overdeflected, and the nut over the platform would be backed off or completely removed. This older design is now generally superseded by precompressed hangers that eliminate the platform and still accomplish the stable position during installation. Note carefully that the number of clean threads when compared with the rusted threads will indicate well over 1 in of deflection in the spring.

This installation actually occurs in a high-probability earthquake zone. Because earthquake shocks can rather easily fracture the platform in this type of isolator, they are not recommended where seismic protection is required.

The neoprene grommet that extends through the hole in the platform plate is also fractured because of excessive weight and misalignment.

FIG. 7.55　Position hanger.

7.56 **All-Directional Seismic Snubbers**
7.57 **and Snubber Tests** ━━━━━━━━━━━━
7.58 Figure 7.56 shows three typical all-directional seismic
7.59 snubbers. The smallest, at the bottom right, has a rated
7.60 capacity of 350 lb. The one on the lower left has a rated
7.61 capacity of 1300 lb. The center snubber, the largest
7.62 standard snubber now produced, has a rated capacity of
7.63 25,000 lb. (See Figure 7.63 for details of design and
construction.)

Figures 7.57 to 7.60 show some of the distinct testing
that seismic snubbing devices should be put through to
verify their paper study capacities and their overload or
factor of safety. Note that the resilient collars are subject
to large deformations. These collars can only withstand
such deformations because they are produced with
bridge-bearing neoprene compounds.

FIG. 7.56   All-directional seismic restraints.

FIG. 7.57   Snubber under shear test.

FIG. 7.58   Snubber under compressor test.

FIG. 7.59   Snubber under bending test.

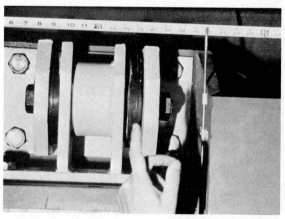

FIG. 7.60   Snubber allowed to resume original position
(refer to Smith-Emery test report in Appendix C).

In Figure 7.60, the hand points to the neoprene element, which has returned to its original shape and size with no apparent evidence of permanent set and no cracks.

Other types of seismic restraints and typical alternatives may be seen on Figures 7.61 and 7.62.

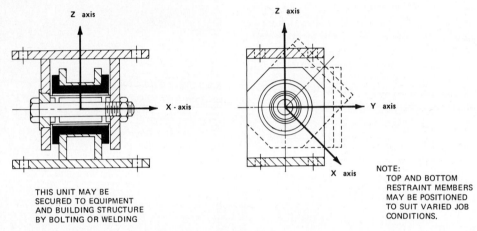

THIS UNIT MAY BE
SECURED TO EQUIPMENT
AND BUILDING STRUCTURE
BY BOLTING OR WELDING

NOTE:
TOP AND BOTTOM
RESTRAINT MEMBERS
MAY BE POSITIONED
TO SUIT VARIED JOB
CONDITIONS.

FIG. 7.61   **All-directional restraint.** *(Mason Industries.)*

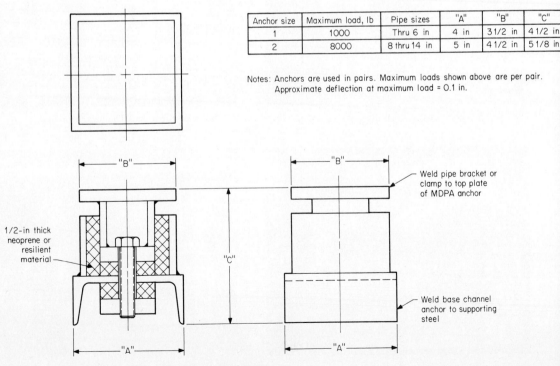

Anchor size	Maximum load, lb	Pipe sizes	"A"	"B"	"C"
1	1000	Thru 6 in	4 in	3 1/2 in	4 1/2 in
2	8000	8 thru 14 in	5 in	4 1/2 in	5 1/8 in

Notes: Anchors are used in pairs. Maximum loads shown above are per pair.
Approximate deflection at maximum load = 0.1 in.

1/2-in thick
neoprene or
resilient
material

Weld pipe bracket or
clamp to top plate
of MDPA anchor

Weld base channel
anchor to supporting
steel

FIG. 7.62   **Resilient pipe anchor unit.** *(Mason Industries.)*

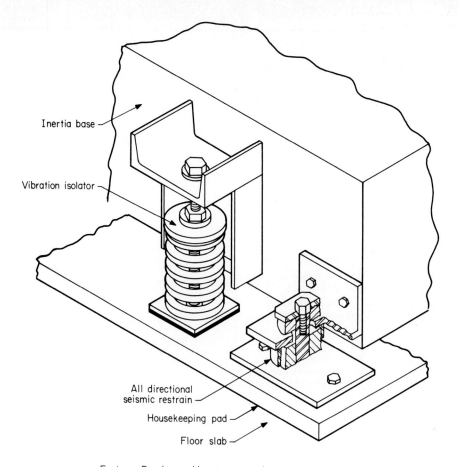

Inertia base

Vibration isolator

All directional
seismic restrain

Housekeeping pad

Floor slab

Feature: Passive snubber type restraint consists of
elastomeric pads and bushings interposed
between welded steel housings.
Mounting configuration: For side mount to equipment base.
Application: For seismic zones 1 and 2 and to some
extent zone 3.

FIG. 7.63   All-directional seismic restraint. *(Korfund Dynamics Corp.)*

7.65
7.66
Figure 7.65 shows an overall view of a gas-engine-driven chiller mounted on massive pocketed, open, stable, steel springs, with rugged seismic protection devices. Pocketed springs were used, rather than the preferred outboard cantilevered type, because of space limitations dictated by the original installation of the concrete housekeeping pad and the reinforcing steel.

In this case, the anchor bolts securing the seismic snubbers to the concrete inertia block extend 3 ft into the block and are then crimped over the internal reinforcing steel with a J hook. Usually, in the floor construction of buildings, it is easier to get to the reinforcing steel without such deep penetrations.

Pull-out data for embedded steel rods and structural capability for bolting to building construction may be found in the *Concrete Reinforcing Institute Handbook*.*

Figure 7.66 provides a close-up view of one of the spring isolators and an adjacent seismic snubber.

The offset placement (see arrow) of the seismic snubber on its steel base plate in Figure 7.64 occurred

---

* See "KSM Structural Engineering Aspects of Headed Concrete Anchors and Deformed Bars in the Concrete Industry," in *Concrete Reinforcing Institute Handbook*, Omar Industries, Schonaumberg, Ill., 1982.

FIG. 7.64   **Securement for seismic restraint.**

FIG. 7.65   Large equipment isolated and seismically restrained.

FIG. 7.66   Close-up of spring isolators and seismic restraint units.

because the steel base plates were attached to the
construction by poured-in-place bolts that had been
mismatched to bolts in the inertia block. In this instance,
the snubbers were cut off their bases and rewelded back
on to correct the misalignment problem.

Special Equipment

8

VIBRATING CONVEYORS

8.1 Conveyor Systems

8.2 Figure 8.1 provides a good view, from underneath a
8.3 vibrating conveyor, of the double-ended drive motor.
8.4 Two separate sections of the conveyor with an offset sec-
8.5 tion (see Figure 8.2) are isolated on open, stable, steel
springs. Although one might assume from the pictures
that the vibration-isolation system is working effectively
to keep unwanted vibrations from traveling through the
building structure, this is not the case. In fact, vibration
transmission problems were so severe at this installation
that doors, windows, ceiling systems, and walls were
vibrating visibly in office areas of the plant 50 to 70 ft
away from the conveyor system.

The conveyor operation produced so much vibrational
energy that building resonances were excited, and the
natural frequency of the selected isolator, being so close
to the building natural frequency, made adequate
vibration control impossible. In addition, at numerous
points, vibrating conveyor components such as steel
support members were allowed to touch sections of plywood deck seen in Figure 8.2 as well as other building
construction components. All such connections provide
paths by which vibrational energy can travel. Several
sections of the conveyor system had been separated from
the overhead building structure and picked up on
separate steel column stanchions standing on the slab on
grade at the main floor (see Figure 8.3). Little was
accomplished by way of reducing vibration transmission
into the building, because the conveyor system was only
partially separated from the building. Contact occurred
at steel supports and along sides of vibrating conveyor
sections near raised portions of the plywood floor.

Note that no isolation was provided under the stanchion base plates and that the plates themselves are

FIG. 8.1   Looking up at conveyor isolation system
and motor drive unit.

FIG. 8.2   Conveyor isolators seen from above.

FIG. 8.3   Floor stanchion for conveyor support.

bolted rigidly into the floor slab. Such direct paths promote the transmission of vibrations. Thus, even if the entire conveyor system had been placed on unisolated freestanding supports, the vibration problem could still have persisted, causing some annoyance. The reason being that workers are positioned on the same grade slab supporting the conveyor system, and thus excessive vibration easily reaches worker locations along the final process assembly line.

The conveying system (shown in Figures 8.1 to 8.5 also includes a vertical spiral section, the top of which is shown in Figure 8.5. The spiral conveyor section oscillates in a horizontal rotary motion.

The base of this vertical section is hard-bolted to the floor at the four corners of the unit. One such connection may be viewed in Figure 8.4. Even though, at these connections, small pieces of neoprene pads were installed between the foot plate and floor, the pads were rendered useless by the direct bolting method.

Steel vertical columns at the four corners of the spiral tray unit run from top to bottom. One such column is seen in the foreground of Figure 8.4. The column seen in Figures 8.4 and 8.5 rubs and bumps against the upper plywood deck as the tray oscillates. Considerable vibration travels into the floor, which is secured to the building structure, thus contributing to the vibration problem.

This sequence of Figures 8.1 to 8.5 demonstrates that professional assistance should be used for such installations. If consultants had been engaged in the early stages of design, the costs for consulting services and the material and labor costs for installing an accurately designed vibration-isolation system would be far lower than the cost of trying to fix this installation. The installation still poses a serious problem; before it is totally corrected, the building owner will probably spend an additional $50,000. To have such a system engineered acoustically in the beginning would probably have cost around $5000, including the installation.

This example reinforces the point that the trial-and-error method can be very expensive. Among other problems with such installations are complaints from workers, loss of worker efficiency, hastening of fatigue in the supporting structure and the building structure, and the waste of paying for something that does not work.

At least one other difficulty in situations where noise and vibration have been a problem is that some people appear to believe that remedial treatment may be prescribed over the telephone. This myth is not helpful in achieving adequate noise and vibration control.

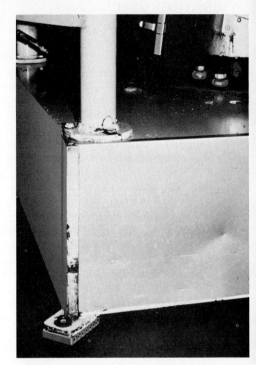

FIG. 8.4 Bottom of spiral tray shaker.

FIG. 8.5 Top of spiral tray shaker.

STAGE LIFT MACHINERY

8.6 Stage Lift Motor and Drive Unit _____

8.7 The piece of mechanical equipment shown in Figure 8.6 is a stage lift motor and drive unit. Generally, this equipment is mounted on its own structural steel base, which is usually rigid enough to allow point-isolating for vibration control, as in this case. The point isolators used here, although not seen installed in the figure, were made up of two layers of ribbed or waffle-pattern neoprene with each layer being about 5/16 in thick.

Even though such equipment is usually in a concrete pit with the pit floor slab on grade, for quietest operation it is best to place such equipment on steel spring vibration isolators. The universal joints and other components of the system that need support from the building structure should also be vibration-isolated.

In some instances, the manufacturer of the equipment may furnish, as standard equipment, some vibration isolators, such as those seen in Figure 8.7. This isolation may or may not be adequate depending on how and to what the total equipment package is mounted.

FIG. 8.7 **Stage lift drive unit.**

FIG. 8.6 **Stage lift machinery.**

ELEVATOR MACHINERY

8.8 Hydraulic Elevator Equipment Isolation,
8.9 Piping, and Piping Isolation

Noise and vibration produced by hydraulic equipment such as that used for elevators and lifts of various types can often adversely affect the acoustical environment in nearby areas. If the hydraulic piping that runs between the pumping unit and the piston passes through noise-sensitive areas, a serious noise problem can arise, even though such spaces may be quite remote from the pumping unit.

Hydraulic piping should be adequately isolated from the building structure in its entire run, including all support points as well as every penetration of the building construction.

Figure 8.8 shows a hydraulic pumping unit in its own mechanical equipment space. The walls of the room have been acoustically treated with fiberglass insulation to aid in reducing the noise reverberation buildup in the mechanical equipment room. Electrical connections to the pump unit have been fitted with sections of flexible conduit to reduce transmission of vibration to the building. The hydraulic pipe has been isolated from the building floor by means of small fabricated steel supports. One such support may be viewed in Figure 8.9, pulled from under the pipe so that its simplicity may be seen. The supports are fabricated from short sections of half-pipe sleeves with flat-plate side supports welded to the half-sleeve. A single layer of ribbed or waffle-pattern neoprene pad is cemented to the inside surface of the half-sleeve, and the hydraulic pipe is then rested on the neoprene. The neoprene selected should be of soft durometer, preferably 30, and the pad should be sized so that it may be loaded properly by the pipe and fluid weight that it must support.

For applications to equipment of the size noted here, the neoprene isolation will generally suffice. However, larger equipment installations will probably require steel spring isolators.

Once the hydraulic pipe runs overhead, it may be isolated with neoprene vibration-isolation hangers, again correctly selected for proper weight loading and static deflection. Penetrations of the building structure should conform with acoustical detail shown in Figure 3.91. The hydraulic pumping unit itself is isolated, in this case, as a complete package on two thicknesses of neoprene pads of 30 durometer. It is sized for proper weight distribution, with a pad mount at each of the four corners of the pumping unit.

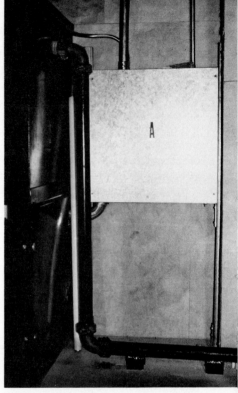

FIG. 8.8   **Piping from hydraulic elevator unit.**

FIG. 8.9 Hydraulic piping isolators.

8.11
8.12

Figures 8.10 to 8.12 show three somewhat different
methods used on the same piping system connected to a
hydraulic elevator pumping unit. None of these arrangements are satisfactory for isolating piping from the
building structure.

Figures 8.10 and 8.11 show the use of very soft foam
rubber as an isolating medium. As you can observe, this
material compresses far too easily, and as the threaded
rod U-bolt is tightened to secure the pipe, the foam
rubber becomes totally displaced.

The neoprene sleeve shown in Figure 8.12 is a better
material to use for isolation, but again, if it is too soft and
overly compressed and not designed properly, it cannot
work to minimize the transmission of vibration.

Using flat bar stock for a strap around the pipe, an
isolating medium will work better to distribute weight
and forces at the securements. In contrast, threaded rod
U-bolts tend to concentrate weight and forces, and the
razor sharp threads easily cut through most resilient
material.

FIG. 8.12   Threaded U-bolt with neoprene sleeve.

FIG. 8.10   Attempt to isolate with foam insulation.

(a)

(b)

FIG. 8.11 *(a)* Foam insulation with threaded U-bolt securement. *(b)* Method of resolving problem shown in *(a)*.

8.13 Motor Generator Set for Elevator

Motor generator sets for elevators may be vibration-isolated in the manner shown in Figure 8.13. Manufactured vibration isolators are employed at the four outrigged feet of this unit. Conduit connections are loose and flexible to limit the amount of vibration transmission via these paths.

Two or three layers of ribbed or waffle-pattern neoprene pads may be used in place of manufactured isolators. Generally, pads should be either 40 or 50 durometer. Pads should be sized so that they are properly weight loaded without exceeding the manufacturer's limitations. A steel plate should be included on top of the pads so that the equipment weight at each loading point is distributed over the entire surface of the end-pad isolator.

Depending on equipment location in relation to noise-sensitive areas, equipment speed and size, and the natural frequency of the supporting structure, steel spring isolators may be required. Thus, for each installation, an engineering assessment should be made.

FIG. 8.13 Motor generator set for elevator.

8.14 Vibration Isolation for Elevator
8.15 Lifting Equipment

Lifting equipment for elevators may be vibration-isolated as required in some noise-sensitive applications as shown in Figures 8.14 and 8.15. Figure 8.14 shows bridge-bearing neoprene pads installed under the feet of the lift, and Figure 8.15 shows the same type of material used in the vertical plane to resist the cable pull from the lift.

Of course, bolting used to hold the lift mechanism in place should be made resiliently as shown in Figure 3.26a.

FIG. 8.15 Resilient snubber for elevator lift equipment.

FIG. 8.14 Bridge-bearing neoprene pads under elevator lift machine.

8.16 Hydraulic Elevator Pumping Unit

Figure 8.16 shows another type of hydraulic pumping equipment package for an elevator or lift. In this case, the entire package may be vibration-isolated as shown on ribbed or waffle-pattern neoprene pads. Preferably two layers of $\frac{5}{16}$- to $\frac{3}{8}$-in-thick layers are generally used, and they must be sized to carry the weight imposed without exceeding the loading limits for the pads.

Piping penetrations through the building construction from the pumping unit to the elevator cylinder must be kept resiliently separated from the building. The method recommended for packing and sealing such penetrations resiliently is shown in Figure 3.91.

FIG. 8.16   Hydraulic elevator pumping unit.

THEATER AND CONCERT HALL WINCHES

8.17 **Vibration Isolation and Noise Control**
8.18 **for Winches** _____

8.19 Stage machinery for theaters, concert halls, and audito-
8.20 riums, where the desired acoustical environment on
8.21 stage and in the hall has to be in the range of NC-15 to
8.22 NC-25, must be considered in terms of the operating
8.23 noise it generates. The general-purpose batten-winch
8.24 machinery shown in Figures 8.17 and 8.18 produces a
8.25 noise measured at about 73 dBA during up-and-down
8.26 operation at a distance of approximately 3 ft from the
8.27 acoustic center of the unit. This winch was under full
8.28 load and operating at maximum speed.
8.29
8.30   The impact noise of the winch brake was measured at
8.31 85 dBA in the up mode of operation and 80 dBA in the
8.32 down mode. Such high noise levels will unfavorably
8.33 impact the stage and house seating areas.
8.34   The brake device can be the single noisiest component
8.35 of winch equipment. The general-purpose batten winch
8.36 shown in Figure 8.18 has its brake mounted on top of the
8.37 drive unit. The impact noise from brake action, whether

FIG. 8.17 Vibration-isolated winch.

FIG. 8.18 **Vibration-isolated batten winch.**

FIG. 8.19 **Captive isolator.**

released or engaged for running the winch either up or down, peaked for the equipment shown in Figure 8.18 at 85 dBA with full load and maximum speed.

Brake disks are another operating noise source. While the brake is being released, the disks vibrate and make a loud chattering sound. It has been recommended to the brake manufacturer that a design feature of the brake should be the application of resilient snubbing elements, which hold loose brake disks steady without interfering with normal brake operation.

Running up and down under full load and at maximum speed, this winch produced a sound level of 73 dBA, at approximately 3 ft (1 m) from the acoustic center of the equipment. The 73-dBA reading does not take into account the impact noise of the brake action at the beginning and end of each run.

This winch generates considerable vibration that can be transmitted to the building structure. Captive isolators were used in this mock-up test assembly, as shown in Figure 8.19. These isolators did not perform satisfactorily during tests, so a different type of isolating device was recommended.

Such equipment needs captive or seismic isolators, because the equipment being isolated not only rests on its support but must pick up substantial positive or negative loads depending on the direction of cable pull. Figure 8.20 shows a typical captive isolator with the internal neoprene components displayed.

Therefore, the cable may tend to lift the winch off its isolating system or increase loading.

FIG. 8.20 **Captive isolator.** *(Mason Industries.)*

A general-purpose batten-winch control console is shown in Figure 8.21 to point out that the console is mechanically ventilated by a smaller propeller fan in the top portion of the assembly (not visible in Figure 8.21), which, while operating, in no way influences the sound level readings measured for the winch equipment.

The winch seen in Figures 8.22 and 8.23 is a type used for raising and lowering adjustable acoustical banners that vary the acoustical environment, usually in a concert hall or similar performing space. This winch is effectively vibration-isolated on neoprene isolators (see Figure 8.23). Normally, the isolators for such winch equipment do not need to be captive, because the banner weight loads imposed on the mounts can be withstood adequately. The measured sound pressure level, A-weighted at a distance of about 3 ft (1 m) from the winch, was 59 to 61 dBA running up and 58 to 63 dBA running down.

The stage machinery manufacturer built an acoustical enclosure for this banner-winch drive assembly. The enclosure was made of plywood, and the entire cover was lined acoustically with 1-in-thick fiberglass with a white vinyl finish face. The two enclosure sections were gasketed with strip black sponge rubber gasket material, which may be seen in Figure 8.24. The only opening in the enclosure was a slightly oversized hole, through which the winch drive shaft passed.

With the acoustical enclosure secured over the winch, as shown in Figure 8.25, the measured sound pressure

FIG. 8.21   Winch console unit.

FIG. 8.22   Banner winch.

FIG. 8.23   Banner-winch isolator.

FIG. 8.24   Acoustic enclosure for winch.

FIG. 8.26   Spot-line winch enclosure.

FIG. 8.25   Winch with acoustic enclosures.

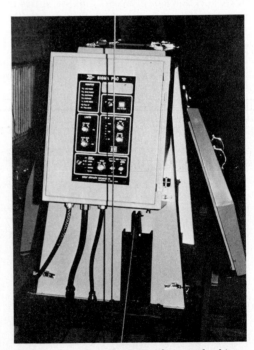

FIG. 8.27   Winch enclosure with control cabinet.

level, A-weighted at a distance of about 3 ft (1 m) from the acoustic center of the drive unit, was 53 dBA running up and 56 dBA running down. Comparison may be made with the unenclosed unit sound measurements above.

The usual practice is not to operate banner winches during performances. Therefore, the acoustical enclosure for the winch might not be needed.

Figure 8.26 shows the front side of acoustically designed covers for two spot-line winches standing side by side. In Figure 8.27 may be seen the control console for one of the spot-line winches. Note that the acoustical door on the right-hand side of the acoustical enclosure has been removed. Each spot-line winch has two such removable sides, either right and left or back and front.

The spot-line winch unit may be seen within its acoustical enclosure but with the covers removed in Figure 8.28, which we shall call the front, and in Figure 8.29, which is the back. Each winch is vibration-isolated within its acoustical box. Note in the foreground of Figure 8.29 the section of flexible electrical conductor, placed so that the winch vibration-isolation system is not bypassed by a solid electrical conduit connection.

These spot-line winches were designed to be portable. Thus, Figure 8.30 shows the type of hold-down device used to secure each unit once it is in place on the channel-iron grating system used in the project.

The control console for each spot-line winch is mounted resiliently to the exterior of the acoustical enclosure. Small neoprene elements were employed for isolation, two at the top corners (see arrows) of each console (Figure 8.31) and two at the bottom corners (Figure 8.32) between the bottom of the console cabinet and the angle-iron shelf bracket. The isolators, not clearly visible in Figure 8.32, are just under the sides of the console cabinet (see arrow). The control console itself is not a noise source, but it is isolated from the winch enclosure so that its own lightweight housing does not reradiate sound from the winch enclosure.

The vibration isolators for the spot-line winch, placed inside the acoustical enclosure, may be seen in Figure 8.33. The neoprene isolator shown was used in tandem with another mount of the same design and characteristics, placed upside down directly underneath the top mount, against the underside of the turned-down steel base plate (see arrow). Thus, the bottom isolator cannot be seen in the figure.

The purpose of the tandem isolator arrangement is to keep the winch vibration-isolated whether it is loaded or unloaded. Like the general-purpose winch, this type of winch must at times pick up substantial loads. When the winch becomes loaded, it tends to lift off its base.

FIG. 8.28   Spot-line winch (front view).

FIG. 8.29   Spot-line winch (back view).

FIG. 8.30   Winch hold-down devices.

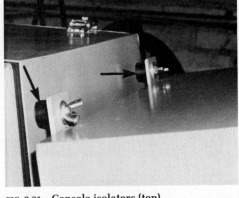

FIG. 8.31   Console isolators (top).

FIG. 8.32   Console isolators (bottom).

FIG. 8.33   Spot-line winch isolators.

FIG. 8.34   Spot-line winch isolators.

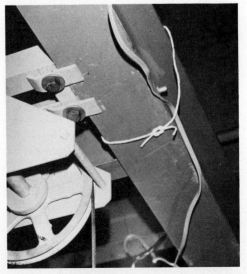

FIG. 8.35   Winch blocks.

Therefore, the underbase isolation mount takes control, and the winch, even while lifting and lowering loads, remains resiliently isolated from the acoustical enclosure base, which is rigidly locked to the grating system.

Further to restrict the movement of the winch while it operates under load, resilient neoprene snubbing elements (see arrow) are placed horizontally at four corners of each winch assembly, as shown in Figure 8.34.

Figure 8.35 shows a head block, and Figures 8.36 and 8.37 are loft blocks, all associated with a general-purpose batten-winch assembly. No vibration isolation is used with the block units to isolate them from the support structure to which they are attached.

During operation and while sound level measurements were being conducted for this test setup, no significant or predominant noise levels emanated from the block assemblies or from the running action of pulleys and cables. However, this finding should not be taken as a guarantee that noise or vibration from such devices will never be a problem.

In the event that head and loft or other types of pulley blocks cause noise problems, the careful installation of neoprene pads and resiliently isolated bolting attachments will probably be effective.

Shown in the following table are the octave band sound-pressure-level readings taken at a location 3 ft (1 m) from the acoustic center of a single operating spot-line winch, with the acoustical enclosure and covers secured in place, with no load and at speed setting 1, which is the slowest speed attainable.

FIG. 8.36   Winch loft block.

FIG. 8.37   Winch head block.

Frequencies, Hz	31	63	125	250	500	1000	2000	4000	8000
SPL decibels up	52	51	49	63	56	45	36	36	43
SPL decibels down	53	54	52	64	57/60	41	35	35	28

To check the impact of the winch operational noise on the acoustical environment in the theater where the tests were being conducted, sound level measurements were made in the theater seating area about 80 ft from the spot-line winch. Background sound levels in the theater with all equipment off were about NC-20. With the winch operating, about an NC-35 was measured. Recommendations were then made to the stage machinery consultants to add a layer of 1 lb/ft² lead backed with ½-in-thick foam rubber on the inside of the enclosure and then to reapply the 1-in fiberglass lining for sound absorption within the box.

The sound transmission loss capability of the enclosure was improved significantly — about 15 dB at 500 Hz.

BEVERAGE COOLERS

8.38 **Vibration Isolation and Noise Control**
8.39 **for Beverage Cooler** ────────────────────────
8.40 Small cooling or refrigeration equipment can pose a
8.41 problem when located in a building such as a concert
8.42 hall where the acoustical design is very critical. The
8.43 noise criterion for most concert halls is NC-15. This level
is very quiet, and almost any type of mechanical equipment produced will easily exceed the NC-15 decibel
levels when operating.

The measured noise levels shown in Figure 8.38 were
made in a hastily produced test chamber. The test
chamber was an 8-ft-square by 8-ft-high plywood box
used to simulate reverberation chamber conditions.
Walls were fabricated with an exterior ½-in-thick
plywood skin, approximately 2 in of styrofoam insulation,
and an inner ½-in-thick plywood skin. The roof of the
box was the same as the walls except that 3½ in of
styrofoam was used between the plywood skins. The box
was sitting on a concrete floor slab on grade. The intake
and exhaust air openings in the roof were temporarily
sealed. The double hollow-core wooden access doors
were loosely gasketed.

The refrigeration unit tested was a CO_2 compressor
cooler for one of the bars in a concert hall. The equipment package includes a compressor, recirculation
pump, carbonator pump, and a small air-circulation fan.
All unit components were mounted on a common steel
base plate. The compressor was vibration-isolated on
rather soft neoprene or rubber isolators located under the
four compressor feet. The recirculation and carbonation
pumps were bolted directly to the unit steel base plate.
The ventilation fan was directly secured to a steel side
panel that connects to the common steel base plate.

Figures 8.39 to 8.42 show a rather small liquid cooling
unit. Although such operating equipment is relatively
quiet, when installed near an acoustically critical space
such as a concert hall, the control of even such a small
source of noise and vibration is an important consideration. Even low noise levels can exceed NC-15.* The
actual sound pressure levels measured for this test
specimen are shown in Figure 8.38. They may be compared with the curve in the noise criteria charts for
NC-15 or other curves. The equipment package includes
specifically a 1-hp compressor, a ¼-hp recirculation
pump, a ¼-hp carbonation pump, and a 1.5-FLA (full

─────────────
* See the noise criteria (NC) graph (Figure 1.1) for decibel
levels at center-frequency octave bands and compare with actual
sound pressure levels measured for this equipment while
operating.

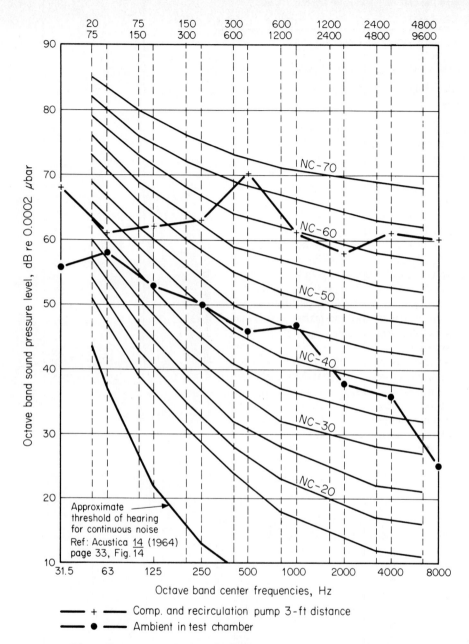

FIG. 8.38 CO₂ cooler noise. Noise criteria (NC) curves. *(From Acustica 14:33 (1964), Figure 14.)*

load amps) ventilation fan. See Figure 8.43 for a manufacturer's schematic diagram.

Sound pressure levels are shown when the cooling unit was operating, measured approximately 3 ft from the equipment acoustic center in an 8-ft-square by 8-ft-high hard materials (wood and concrete) test chamber.

	Octave Band Center Frequency								
	31	63	125	250	500	1000	2000	4000	8000
SPL, dB	68	61	65	63	70	61	58	61	60

In the actual project installation, the CO_2 coolers were located in small separate equipment rooms. Vibration isolation for the entire package can be in the form of ribbed or waffle-pattern neoprene pads or manufactured rubber-in-shear mountings. In either case, care should be taken to not exceed the manufacturer's weight-loading requirements for the given isolator, and the durometer of the isolating material should be in the range of 30 to 40.

Depending on the location of this type of equipment in relation to adjacent public or noise-sensitive spaces, it may be desirable to gasket the door to the space in which these units are located. The door itself may have to be of a particular sound rating class (STC). Such requirements for the door will help to ensure that the desired acoustical environment just outside the CO_2 cooler equipment space is not compromised.

Figure 8.38 shows the exterior casing, which houses the compressor unit and the small vent fan. The carbonator tank is located just to the right of the unit casing. Figure 8.40 shows the carbonator pump (foreground) and the recirculation pump (background). Note that both pumps are bolted directly to the unit's common base

FIG. 8.40   Beverage cooler interior.

FIG. 8.39   Beverage cooler.

without resilient vibration isolation. You can look into
the compressor housing in Figure 8.41, and Figure 8.42 is
a view from closer in, so that the neoprene vibration
isolators under one of the compressor feet are discernible
(see arrow).

FIG. 8.41   Beverage cooler.

FIG. 8.42   Beverage cooler isolator.

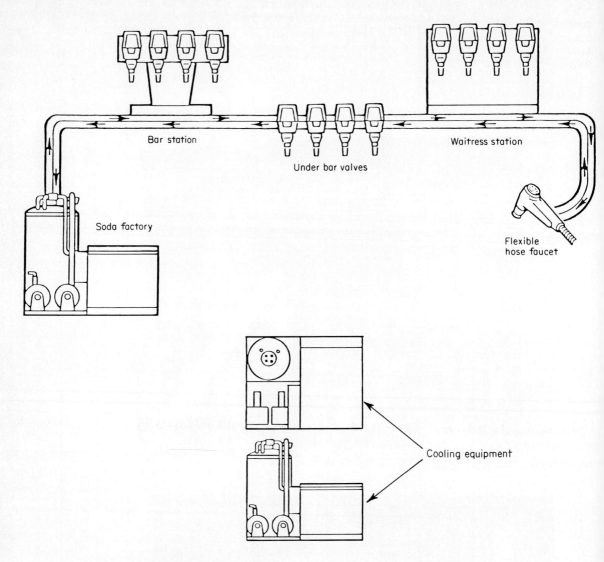

Bar station

Under bar valves

Waitress station

Soda factory

Flexible
hose faucet

Cooling equipment

FIG. 8.43 Schematic diagram for CO_2 cooler system.

ESCALATORS

8.44 Escalator Drive Unit ————————————————————————

8.45 A typical escalator is shown in Figure 8.44. There are
8.46 many moving parts in an escalator, all of them powered
8.47 by a drive unit that is located somewhere in the escalator module. A typical basic drive unit is shown in Figure
8.45. Highlighted in Figure 8.45 are the vibration isolators that are furnished as standard equipment with such
units. For a typical simplified plan and elevation of an
escalator, see Figure 8.46. An actual drive unit may be
seen in Figure 8.47.

Escalator manufacturers have developed a modular
concept that allows them to join almost any number of
modules together in series to form a continuous escalator.

Figure 8.47 shows a motor-drive unit (see arrow) of an
escalator. Locations of drive units within an escalator
section will vary with the manufacturer. This escalator
has its drive unit at the top; others may occur in the
middle or at the bottom of the escalator section.

In the application shown in Figure 8.47, the entire
escalator section including the drive unit was vibration
isolated at the top and bottom support points that rest on
the building structure. The isolation used was ½-in-thick
pads of bridge-bearing neoprene, sized to be adequately
loaded for achieving the proper deflection. This particular
drive unit is not vibration-isolated from the escalator
section itself.

FIG. 8.44   Escalator.

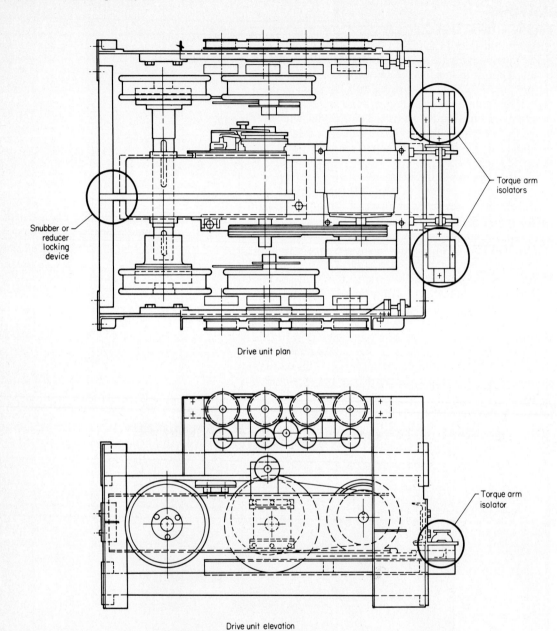

Snubber or
reducer
locking
device

Torque arm
isolators

Drive unit plan

Torque arm
isolator

Drive unit elevation

FIG. 8.45   **Escalator drive unit drawing.** *(After Westinghouse.)*

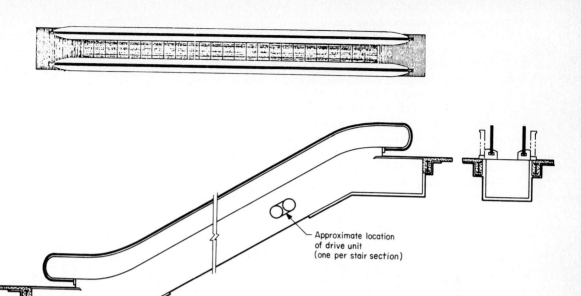

FIG. 8.46   Escalator. *(After Westinghouse.)*

Approximate location
of drive unit
(one per stair section)

FIG. 8.47   Escalator drive unit installed.

8.50 Although escalators do make noise (Figure 8.48) and
8.51 produce vibrations (Figure 8.49), the manufacturers have
done a rather good job of noise control. Figure 8.48
shows noise levels actually measured on and around an
escalator installed in a bank building. No complaints of
excessive noise or vibration have been registered by the
building owners. These measurements were made with
the encouragement of the escalator manufacturer as part
of a survey to verify the claim of low noise levels. The
claims were substantiated.

The drive unit electric motor of an escalator (pulleys,
sheaves, and gear reducer) generally has two resilient
torque arm isolators and one resilient snubber called a
reducer locking device (see Figure 8.45). These units
cushion and somewhat limit the amount of movement of
the drive unit caused by starting torque. Their function
is not primarily as vibration-isolation devices, but they do
reduce energy transmission from the drive unit to the
escalator frame. At other points, however, the drive unit
is bolted rigidly to the escalator frame.

The stair assembly of the escalator rides on resilient
rollers manufactured from polyurethane. The stair
assembly horizontal and vertical guide rollers are also
resilient polyurethane.

The chain drive for this escalator is fabricated from
laminated metal sections. The engagement of the chain is
by polyurethane teeth set in the drive sprocket, so that
the metal parts of the two components do not touch each
other.

The electric motor drive is generally about a 10-hp
motor, running at 1745 r/min.

The gear reducers in this escalator are of link-belt
manufacture and are initially arranged to drive the stair
at 90 ft/min. Speed changes are made by changing motor
and/or drive sheaves and belts.

The term "modular" is used for this model to designate
that 20-ft straight sections, each with its drive unit, may
be added almost indefinitely to extend the overall stair
length to suit various building configurations.

The stairs may be shipped totally shop-fabricated
except for the structural sides and the handrails, which
are attached to the stair at the jobsite. The main reason
for this shipping arrangement is to reduce breakage of
the structural glass elements.

The handrails are rubber and are molded around
layers of canvas and a stainless-steel tape to form one
continuous element. Then handrails are driven by
friction wheels usually made of polyurethane.

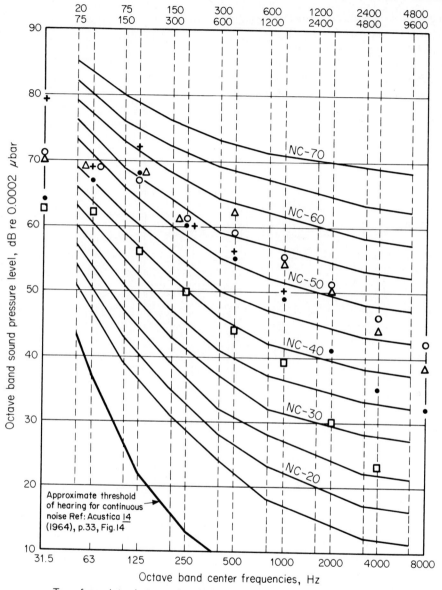

- • Top of escalator between handrails
- + Under escalator 12 in from drive enclosure
- o Up escalator only operating - 4 ft above drive
- △ Down escalator only operating - 4 ft above drive
- □ Background (escalator off)

FIG. 8.48   Escalator noise levels. *(Westinghouse model modular 32-in escalator.)*

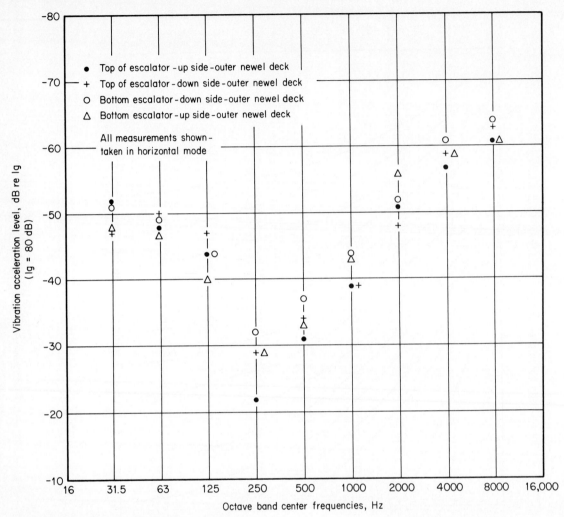

FIG. 8.49   **Escalator vibration levels.**

Each stair tread is connected to the stair assembly separately. At the two connecting points for each tread, neoprene or rubber bushings are provided so that the metal parts do not touch, and thus little vibration occurs.

The trail rollers (two) for each individual stair tread are of polyurethane construction, so that each tread is resiliently isolated from the stair frame and running mechanisms.

Octave frequency band, Hz	SPL in octave band, dB	A-scale correction term, dB	Corrected band, dB	dB Addition* of corrected values
31	71	−39	32	
				44
63	69	−26	43	
				52
125	67	−16	51	
				55
250	61	−9	52	
				58 — 61 dBA
500	59	−3	56	
1000	55	0	55	
				57
2000	51	+1	52	
				53
4000	46	+1	47	
				48
8000	43	−1	42	

FIG. 8.50 Octave band sound-pressure-level measurements for a modular 32-in escalator.

These measured sound pressure levels are listed in Figure 8.50, and A-scale correction terms have been applied in order to add the octave band sound pressure levels by decibel addition to arrive at a single number dBA rating. Of course, the A-scale reading may be taken in the field or test facility with a sound level meter. It is helpful to know the method for converting to dBA when one has only sound pressure levels in octave band center frequencies.

Decibel addition. Decibels are logarithmic quantities and do not follow normal algebraic rules for addition. Instead, decibels are first converted to energy equivalents, the energy equivalents are added algebraically, and then the total energy equivalent is converted back to its

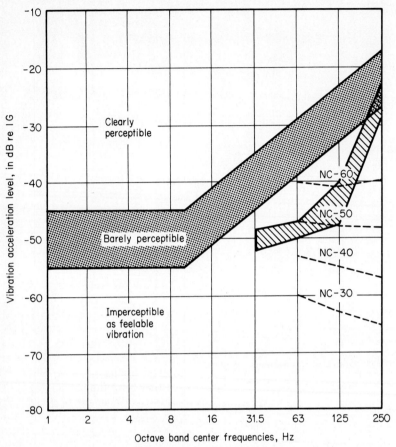

FIG. 8.51 Vibration-acceleration levels near the onset of "feelability."

decibel value. There a number of simplifying methods for carrying out this procedure. One method is shown here.

Procedures for Quick Addition of Sound Levels or Other Decibel Quantities by "Decibel Addition"
For adding any two decibel levels to an accuracy of about 1 dB

When Two Decibel Values Differ by	Add the Following Amount to the Higher Value
0 or 1 dB	3 dB
2 or 3 dB	2 dB
4–9 dB	1 dB
10 dB or more	0 dB

NOTE: When adding several levels, start with lowest levels first: continue two at a time until only one final value remains.

Figure 8.51 shows vibration-acceleration levels near the onset of "feelability."

Wire Rope Identification and Construction*

Wire rope is identified not only by its component parts but also by its construction, i.e., by the way the wires have been laid to form strands and by the way the strands have been laid around the core.

Figure A.1a and c shows strands as normally laid into the rope to the right—in a fashion similar to the threading in a right-hand bolt. Conversely, the "left lay" rope strands (Figure A.1b and d) are laid in the opposite direction.

Again, Figure A.1a and b shows *regular lay* ropes. Following these are the types known as *lang lay* ropes. Note that the wires in regular lay ropes appear to line up with the axis of the rope; in lang lay rope, the wires form an angle with the axis of the rope. This difference in appearance is a result of variations in manufacturing techniques: regular lay ropes are made so that the direction of the wire lay in the strand is opposite to the direction of the strand lay in the rope; lang lay ropes (Figure A.1c and d) are made with both strand lay and rope lay in the same direction. Finally, the type in Figure A.1e, called *alternate lay*, consists of alternating regular and lang lay strands.

Of all wire rope types in current use, *right regular lay* is found in the widest range of applica-

tions. Many applications related to excavation, construction, or mining require *lang lay* rope. Currently, *left lay* rope is used less frequently. In any case, where left lay and/or lang lay are required, the manufacturer/supplier must be so informed before ordering. As for *alternate lay* ropes, these are used for special applications.

Circumstances that favor the use of lang lay ropes derive from two unique advantages over regular lay ropes. Lang lay ropes (1) are more resistant to bending fatigue and (2) have a greater wearing surface per wire across the crown of the strand. The total wearing surface area of the rope is, for practical purposes, the same for both regular and lang lay ropes—with the same geometric construction and depth of wear—the eventual wear on the equipment and the service life of the rope favors lang lay construction on applications where fatigue or abrasion are controlling factors.

To illustrate this point, Figure A.2 compares a regular lay with a lang lay rope, each of which has been worn to the same amount of reduction in their respective diameters.

Hence, it is not the total of the rope's worn surface area that governs the life span of rope and equipment. It is, rather, the inherent characteristics of properly used lang lay ropes that gives them a significant advantage in resistance to both abrasion and fatigue.

---

* This appendix is adapted from the *Wire Rope Users Manual*, American Iron and Steel Institute, Washington, D.C., 1979.

FIG. A.1 A comparison of typical wire rope lays: *(a)* right regular lay, *(b)* left regular lay, *(c)* right lang lay, *(d)* left lang lay, *(e)* right alternate lay.

FIG. A.2 A comparison of wear characteristics between *lang lay* and *regular lay* ropes. The line a-b indicates the rope axis.

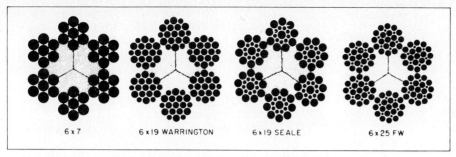

FIG. A.3   Basic constructions around which standard wire ropes are built.

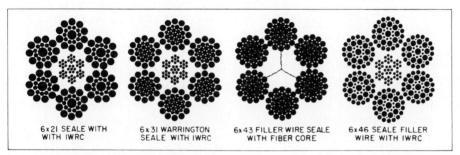

FIG. A.4   A few combinations of basic design constructions.

However, lang lay ropes have some disadvantages. They are more susceptible to damage resulting from handling abuses, bending over extremely small sheaves, pinching in undersize sheave grooves, and crushing when improperly wound on drums, and they are subject to excessive rotation. In fact, this latter tendency for the rope and the strands to unwind in the same direction requires that lang lay ropes should be secured at both ends to prevent unlaying or spinning out.

Preforming is a wire rope manufacturing process wherein the strands and their wires are shaped—during fabrication—to the spiral form that they will ultimately assume in the finished rope or strand.

As previously noted, wire rope strands are made up of a number of wires. The wire arrangement in the strands will determine the rope's functional characteristics, i.e., its capacity to meet the operating conditions to which it will be subjected. There are many basic design constructions around which standard wire ropes are built; some of these are shown in Figure A.3.

Four typical strand cross sections, designed around the Warrington, Seale, and filler wire basic constructions are shown in Figure A.4.

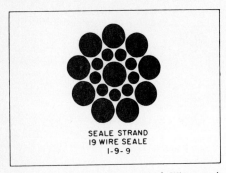

SEALE STRAND
19 WIRE SEALE
1-9-9

FIG. A.5 A single wire rope strand. Wire rope is identified by reference to its number of strands as well as the number and geometric arrangement of wires in the strand.

Wire ropes are identified by a nomenclature that is referenced to (1) the number of strands in the rope, (2) the number (nominal or exact) and arrangement of wires in each strand, and (3) a descriptive word or letter indicating the type of construction, i.e., geometric arrangement of wires

(Fig. A.5). The most widely used classifications are listed and described in Table A.1.

At this point, it may be useful to discuss wire rope nomenclature in somewhat greater detail. It is a subject that may easily generate some misunderstanding. The reason for this stems from the practice of referring to rope either by class or by its specific construction.

Ropes are classified both by the number of strands and the number of wires in each strand, e.g., 6×7, 6×19, 6×37, 8×19, 19×7. However, these are nominal classifications that may or may not represent the actual construction. For example, the 6×19 class commonly includes constructions such as 6×21 filler wire, 6×25 filler wire, and 6×26 Warrington Seale. Despite the fact that none of these have 19 wires, they are designated as being in the 6×19 classification.

Hence, a supplier receiving an order for 6×19 rope may assume this to be a *class* reference and is legally justified in furnishing any construction within this category. But, if the job should require

Table A.1   Wire Rope Classifications
Based on the nominal number of wires in each strand

Classification	Description
6×7	Containing 6 strands that are made up of 3 through 14 wires, of which no more than 9 are outside wires.
6×19	Containing 6 strands that are made up of 15 through 26 wires, of which no more than 12 are outside wires.
6×37	Containing 6 strands that are made up of 27 through 49 wires, of which no more than 18 are outside wires.
6×61	Containing 6 strands that are made up of 50 through 74 wires, of which no more than 24 are outside wires.
6×91	Containing 6 strands that are made up of 75 through 109 wires, of which no more than 30 are outside wires.
6×127	Containing 6 strands that are made up of 110 or more wires, of which no more than 36 are outside wires.
8×19	Containing 8 strands that are made up of 15 through 26 wires, of which no more than 12 are outside wires.
19×7 and 18×7	Containing 19 strands, each strand is made up of 7 wires. It is manufactured by covering an inner rope of 7×7 left lang lay construction with 12 strands in right regular lay. (The rotation-resistant property that characterizes this highly specialized construction is a result of the counter torques developed by the two layers.) When the steel wire core strand is replaced by a fiber core, the description becomes 18×7.

Table A.2 Terminal Efficiencies (Approximate)
Efficiencies are based on nominal strengths

Method of Attachment	Efficiency, %	
	Rope with IWRC*	Rope with FC†
Wire rope socket-spelter or resin attachment	100	100
Swaged socket	95	(Not established)
Mechanical spliced sleeve		
1 in diameter and smaller	95	92½
1⅛ in through 1⅞ in diameter	92½	90
2 in diameter and larger	90	87½
Loop or thimble splice—hand spliced (tucked) (carbon steel rope)		
¼ in	90	90
5/16 in	89	89
⅜ in	88	88
7/16 in	87	87
½ in	86	86
⅝ in	84	84
¾ in	82	82
⅞ through 2½ in	80	80
Loop or thimble splice—hand spliced (tucked) (stainless steel rope)		
¼ in	80	
5/16 in	79	
⅜ in	78	
7/16 in	77	
½ in	76	
⅝ in	74	
¾ in	72	
⅞ in	70	
Wedge sockets‡ (depending on design)	75 to 90	75 to 90
Clips‡ (number of clips varies with size of rope)	80	80

* IWRC = independent wire rope core.
† FC = fiber core.
‡ Typical values when applied properly. Refer to fittings manufacturers for exact values and method.

the special characteristics of 6 × 25 W, and a 6 × 19 Seale (Figure A.3) is supplied in its stead, a shorter service life can be expected.

To avoid such misunderstanding, the safest procedure is to order a specific construction if such geometry is essential for the intended purpose or to order both by class and construction, e.g., 6 × 19 (6 × 26 Warrington Seale).

Identifying wire rope in *class* groups facilitates selection on the basis of strength, weight per foot, and price since all ropes within a class have the same nominal strength, weight per foot, and price. As for other functional characteristics, these can be obtained by referencing the specific construction within the class.

Only three wire ropes in the 6 × 19 classification actually have 19 wires: 6 × 19 two operation, 6 × 19 Seale, and 6 × 19 Warrington. All the rest

have different counts. There is a greater proportion of 37-wire constructions in the 6 × 37 class, but these are infrequently produced. The more commonly available 6 × 37 constructions include 6 × 31 Seale, 6 × 31 Warrington Seale (WS), 6 × 36 WS, 6 × 41 Seale filler wire (SFW), 6 × 41 WS, 6 × 43 FW, and 6 × 46 WS—none of which contain 37 wires.

While a strand's interior has some significance, its important characteristics relate to the number

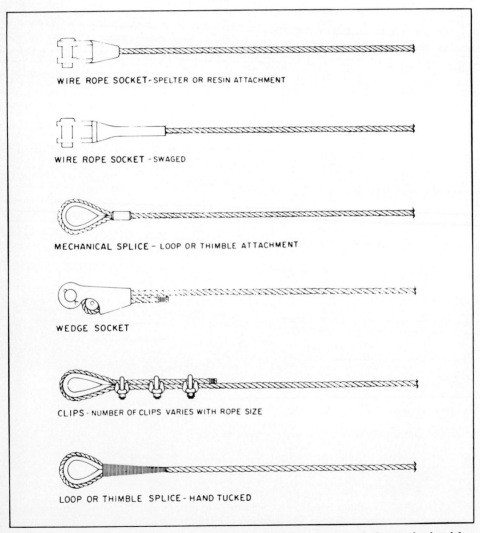

WIRE ROPE SOCKET - SPELTER OR RESIN ATTACHMENT

WIRE ROPE SOCKET - SWAGED

MECHANICAL SPLICE – LOOP OR THIMBLE ATTACHMENT

WEDGE SOCKET

CLIPS - NUMBER OF CLIPS VARIES WITH ROPE SIZE

LOOP OR THIMBLE SPLICE - HAND TUCKED

FIG. A.6 End fittings, or attachments, are available in many designs, some of which were developed for particular applications. The six shown are among the most commonly used.

and, in consequence, the size of the outer wires.

Wire rope nomenclature also defines length, size (i.e., diameter), type, direction of lay, grade of rope, type of core, and whether it is preformed (p/f) or nonpreformed (np/f). If the direction and type of lay are omitted from the rope description, it is presumed to be a right regular lay. In addition, if no mention is made as to preforming, this will be presumed as a requirement for preforming. On the other hand, an order for elevator rope requires an explicit statement since p/f and np/f ropes are used extensively.

An example of a complete description would appear thus:

> 600 ft   ¾ in   6 × 25 FW *left lang lay*
> Improved plow   IWRC

(Rope described above would be made *preformed*.)

When a center wire is replaced by a strand, it is considered as a single wire, and the rope classification remains unchanged.

There are, of course, many other types of wire rope, but they are useful only in a limited number of applications and, as such, are sold as specialties (Figure A.6).

WIRE ROPE CLIPS

Wire rope clips are widely used for attaching wire rope to haulages, mine cars, and hoists and for joining two ropes.

Clips are available in two basic designs; the *U bolt* and *fist grip* (Figure A.7). The efficiency of both types is the same.

When using U-bolt clips, extreme care must be exercised to make certain that they are attached correctly, i.e., the U bolt must be applied so that the "U" section is in contact with the dead end of the

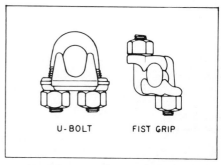

FIG. A.7   Wire rope clips are obtainable in two basic designs: *U bolt* and *fist grip.* Their efficiency is the same.

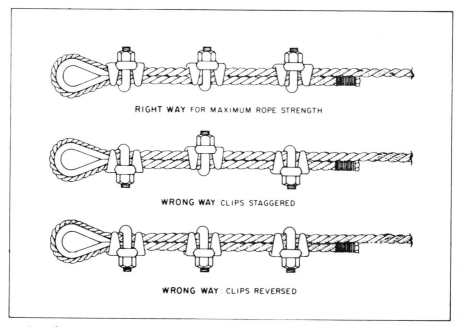

RIGHT WAY FOR MAXIMUM ROPE STRENGTH

WRONG WAY: CLIPS STAGGERED

WRONG WAY: CLIPS REVERSED

FIG. A.8   The *correct* way to attach U bolts is shown at the top; the U section is in contact with the rope's dead end.

rope (Figure A.8). Also, the tightening and retightening of the nuts must be accomplished as required.

HOW TO APPLY U-BOLT CLIPS (Table A.3)

Recommended method of applying U-bolt clips to get maximum holding power of the clip:

1. Turn back the specified amount of rope from the thimble. Apply the first clip one base width from the dead end of the wire rope (U bolt over dead end — live end rests in clip saddle). Tighten nuts evenly to recommended torque.

2. Apply the next clip as near the loop as possible. Turn on nuts firm but do not tighten.

3. Space additional clips if required equally

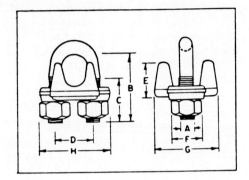

Table A.3*

Clip Size	A	B	C	D	E	F	G	H	Minimum number of clips	Amount of rope to turn back	Torque, lb/ft	Weight, lb/100
1/8	0.22	0.72	0.44	0.47	0.41	0.38	0.81	0.94	2	3¼	4.5	5
3/16	0.25	0.97	0.56	0.59	0.50	0.44	0.94	1.16	2	3¾	7.5	9
1/4	0.31	1.03	0.50	0.75	0.66	0.56	1.19	1.44	2	4¾	15	18
5/16	0.38	1.38	0.75	0.88	0.72	0.69	1.31	1.69	2	5¼	30	30
3/8	0.44	1.50	0.75	1.00	0.91	0.75	1.63	1.94	2	6½	45	42
7/16	0.50	1.88	1.00	1.19	1.03	0.88	1.81	2.28	2	7	65	70
1/2	0.50	1.88	1.00	1.19	1.13	0.88	1.91	2.28	3	11½	65	75
9/16	0.56	2.25	1.25	1.31	1.22	0.94	2.06	2.50	3	12	95	100
5/8	0.56	2.38	1.25	1.31	1.34	.94	2.06	2.50	3	12	95	100
3/4	0.63	2.75	1.44	1.50	1.41	1.06	2.25	2.84	4	18	130	150
7/8	0.75	3.13	1.63	1.75	1.59	1.25	2.44	3.16	4	19	225	240
1	0.75	3.50	1.81	1.88	1.78	1.25	2.63	3.47	5	26	225	250
1 1/8	0.75	3.88	2.00	2.00	1.91	1.25	2.81	3.59	6	34	225	310
1 1/4	0.88	4.25	2.13	2.31	2.19	1.44	3.13	4.13	6	37	360	460
1 3/8	0.88	4.63	2.31	2.38	2.31	1.44	3.13	4.19	7	44	360	520
1 1/2	0.88	4.94	2.38	2.59	2.53	1.44	3.41	4.44	7	48	360	590
1 5/8	1.00	5.31	2.63	2.75	2.66	1.63	3.63	4.75	7	51	430	730
1 3/4	1.13	5.75	2.75	3.06	2.94	1.81	3.81	5.28	7	53	590	980
2	1.25	6.44	3.00	3.38	3.28	2.00	4.44	5.88	8	71	750	1340
2 1/4	1.25	7.13	3.19	3.88	3.94	2.00	4.56	6.38	8	73	750	1570
2 1/2	1.25	7.69	3.44	4.13	4.44	2.00	4.69	6.63	9	84	750	1790
2 3/4	1.25	8.31	3.56	4.38	4.88	2.00	5.00	6.88	10	100	750	2200
3	1.50	9.19	3.88	4.75	5.34	2.38	5.31	7.63	10	106	1200	3200

* From The Crosby Group.

Table A.4 Nominal Strengths of Wire Rope
6 × 7 Classification/bright (uncoated), fiber core

Nominal Diameter		Approximate Mass		Nominal Strength* Improved Plow Steel	
in	mm	lb/ft	kg/m	tons	t
¼	6.5	0.09	0.14	2.64	2.4
⁵⁄₁₆	8	0.15	0.22	4.10	3.72
⅜	9.5	0.21	0.31	5.86	5.32
⁷⁄₁₆	11.5	0.29	0.43	7.93	7.2
½	13	0.38	0.57	10.3	9.35
⁹⁄₁₆	14.5	0.48	0.71	13.0	11.8
⅝	16	0.59	0.88	15.9	14.4
¾	19	0.84	1.25	22.7	20.6
⅞	22	1.15	1.71	30.7	27.9
1	26	1.50	2.23	39.7	36.0
1⅛	29	1.90	2.83	49.8	45.2
1¼	32	2.34	3.48	61.0	55.3
1⅜	35	2.82	4.23	73.1	66.3
1½	38	3.38	5.03	86.2	78.2

* To convert to kilonewtons (kN), multiply tons (nominal breaking strength) by 8.896; 1 lb = 4.448 N.

Table A.5 Nominal Strengths of Wire Rope
6 × 7 Classification/bright (uncoated), IWRC

Nominal Diameter		Approximate Mass		Nominal Strength* Improved Plow Steel	
in	mm	lb/ft	kg/m	tons	t
¼	6.5	0.10	0.15	2.84	2.58
⁵⁄₁₆	8	0.16	0.24	44.1	4.0
⅜	9.5	0.23	0.34	6.30	5.72
⁷⁄₁₆	11.5	0.32	0.48	8.52	7.73
½	13	0.42	0.63	11.1	10.1
⁹⁄₁₆	14.5	0.53	0.79	14.0	12.7
⅝	16	0.65	0.97	17.1	15.5
¾	19	0.92	1.37	24.4	22.1
⅞	22	1.27	1.89	33.0	29.9
1	26	1.65	2.46	42.7	38.7
1⅛	29	2.09	3.11	53.5	48.5
1¼	32	2.57	3.82	65.6	59.5
1⅜	35	3.12	4.64	78.6	71.3
1½	38	3.72	5.54	92.7	84.1

* To convert to kilonewtons (kN), multiply tons (nominal breaking strength) by 8.896; 1 lb = 4.448 N.

between the first two. Turn on nuts—take up rope slack—tighten all nuts evenly on all clips to recommended torque.

4. *Notice:* Apply the initial load and retighten nuts to the recommended torque. Rope will stretch and shrink in diameter when loads are applied. Inspect periodically and retighten.

A termination made in accordance with the above instructions, and using the number of clips shown, has an approximate 80 percent efficiency rating. This rating is based upon the catalog breaking strength of wire rope. If a pulley is used in place of a thimble for turning back the rope, add one additional clip.

The number of clips shown is based upon using right regular or lang lay wire rope, 6 × 19 class or 6 × 37 class, fiber core or IWRC, IPS, or XIPS. If Seale construction or similar large outer-wire-type construction in the 6 × 19 class is to be used for

sizes 1 in and larger, add one additional clip.

The number of clips shown also applies to right regular lay wire rope, 8 × 19 class, fiber core, IPS, sizes 1½ in and smaller; and right regular lay wire rope, 18 × 7 class, fiber core, IPS or XIPS, sizes 1¾ in and smaller.

For other classes of wire rope not mentioned above, it may be necessary to add additional clips to the number shown.

If a greater number of clips are used than shown in the table, the amount of rope turnback should be increased proportionately. *The above is based on use of clips on new rope.*

Important: Failure to make a termination in accordance with the aforementioned instructions or failure to periodically check and retighten to the recommended torque will cause a reduction in efficiency rating.

Table A.6 Nominal Strengths of Wire Rope
6 × 19 Classification/bright (uncoated), fiber core

Nominal Diameter		Approximate Mass		Nominal Strength* Improved Plow Steel	
in	mm	lb/ft	kg/m	tons	t
¼	6.5	0.11	0.16	2.74	2.49
5/16	8	0.16	0.24	4.26	3.86
⅜	9.5	0.24	0.35	6.10	5.53
7/16	11.5	0.32	0.48	8.27	7.50
½	13	0.42	0.63	10.7	9.71
9/16	14.5	0.53	0.79	13.5	12.2
⅝	16	0.66	0.98	16.7	15.1
¾	19	0.95	1.41	23.8	21.6
⅞	22	1.29	1.92	32.2	29.2
1	26	1.68	2.5	41.8	37.9
1⅛	29	2.13	3.17	52.6	47.7
1¼	32	2.63	3.91	64.6	58.6
1⅜	35	3.18	4.73	77.7	70.5
1½	38	3.78	5.63	92.0	83.5
1⅝	42	4.44	6.61	107	97.1
1¾	45	5.15	7.66	124	112
1⅞	48	5.91	8.8	141	128
2	51	6.72	10.0	160	145
2⅛	54	7.59	11.3	179	162
2¼	57	8.51	12.7	200	181
2⅜	61	9.48	14.1	222	201
2½	64	10.5	15.6	244	221
2⅝	67	11.6	17.3	268	243
2¾	70	12.7	18.9	292	265

* To convert to kilonewtons (kN), multiply tons (nominal breaking strength) by 8.896; 1 lb = 4.448 N.

Table A.7 Nominal Strengths of Wire Rope
6 × 19 Classification/bright (uncoated), IWRC

Nominal Diameter		Approximate Mass		Nominal Strength* Improved Plow Steel		Extra Improved Plow Steel	
in	mm	lb/ft	kg/m	tons	t	tons	t
¼	6.5	0.12	0.17	2.94	2.67	3.40	3.08
5/16	8	0.18	0.27	4.58	4.16	5.27	4.78
⅜	9.5	0.26	0.39	6.56	5.95	7.55	6.85
7/16	11.5	0.35	0.52	8.89	8.07	10.2	9.25
½	13	0.46	0.68	11.5	10.4	13.3	12.1
9/16	14.5	0.59	0.88	14.5	13.2	16.8	15.2
⅝	16	0.72	1.07	17.7	16.2	20.6	18.7
¾	19	1.04	1.55	25.6	23.2	29.4	26.7
⅞	22	1.42	2.11	34.6	31.4	39.8	36.1
1	26	1.85	2.75	44.9	40.7	51.7	46.9
1⅛	29	2.34	3.48	56.6	51.3	65.0	59.0
1¼	32	2.89	4.30	69.4	63.0	79.9	72.5
1⅜	35	3.5	5.21	83.5	75.7	96.0	87.1
1½	38	4.16	6.19	98.9	89.7	114	103
1⅝	42	4.88	7.26	115	104	132	120
1¾	45	5.67	8.44	133	121	153	139
1⅞	48	6.5	9.67	152	138	174	158
2	51	7.39	11.0	172	156	198	180
2⅛	54	8.35	12.4	192	174	221	200
2¼	57	9.36	13.9	215	195	247	224
2⅜	61	10.4	15.5	239	217	274	249
2½	64	11.6	17.3	262	238	302	274
2⅝	67	12.8	19.0	288	261	331	300
2¾	70	14.0	20.8	314	285	361	327

* To convert to kilonewtons (kN), multiply tons (nominal breaking strength) by 8.896; 1 lb = 4.448 N.

Table A.8 Nominal Strengths of Wire Rope
6 × 37 Classification/bright (uncoated), fiber core

Nominal Diameter		Approximate Mass		Nominal Strength* Improved Plow Steel	
in	mm	lb/ft	kg/m	tons	t
¼	6.5	0.11	0.16	2.74	2.49
⁵⁄₁₆	8	0.16	0.24	4.26	3.86
⅜	9.5	0.24	0.35	6.10	5.53
⁷⁄₁₆	11.5	0.32	0.48	8.27	7.50
½	13	0.42	0.63	10.7	9.71
⁹⁄₁₆	14.5	0.53	0.79	13.5	12.2
⅝	16	0.66	0.98	16.7	15.1
¾	19	0.95	1.41	23.8	21.6
⅞	22	1.29	1.92	32.2	29.2
1	26	1.68	2.50	41.8	37.9
1⅛	29	2.13	3.17	52.6	47.7
1¼	32	2.63	3.91	64.6	58.6
1⅜	35	3.18	4.73	77.7	70.5
1½	38	3.78	5.63	92.0	83.5
1⅝	42	4.44	6.61	107	97.1
1¾	45	5.15	7.66	124	112
1⅞	48	5.91	8.8	141	128
2	51	6.72	10.0	160	145
2⅛	54	7.59	11.3	179	162
2¼	57	8.51	12.7	200	181
2⅜	61	9.48	14.1	222	201
2½	64	10.5	15.6	244	221
2⅝	67	11.6	17.3	268	243
2¾	70	12.7	18.9	292	265
2⅞	74	13.9	20.7	317	287
3	77	15.1	22.5	344	312
3⅛	80	16.4	24.4	371	336
3¼	83	17.7	26.3	399	362

* To convert to kilonewtons (kN), multiply tons (nominal breaking strength) by 8.896; 1 lb = 4.448 N.

Table A.9 Nominal Strengths of Wire Rope
6 × 37 Classification/bright (uncoated), IWRC

Nominal Diameter		Approximate Mass		Nominal Strength* Improved Plow Steel		Extra Improved Plow Steel	
in	mm	lb/ft	kg/m	tons	t	tons	t
¼	6.5	0.12	0.17	2.94	2.67	3.4	3.08
⁵⁄₁₆	8	0.18	0.27	4.58	4.16	5.27	4.78
⅜	9.5	0.26	0.39	6.56	5.95	7.55	6.85
⁷⁄₁₆	11.5	0.35	0.52	8.89	8.07	10.2	9.25
½	13	0.46	0.68	11.5	10.4	13.3	12.1
⁹⁄₁₆	14.5	0.59	0.88	14.5	13.2	16.8	15.2
⅝	16	0.72	1.07	17.9	16.2	20.6	18.7
¾	19	1.04	1.55	25.6	23.2	29.4	26.7
⅞	22	1.42	2.11	34.6	31.4	39.5	35.9
1	26	1.85	2.75	44.9	40.7	51.7	46.9
1⅛	29	2.34	3.48	56.5	51.3	65.0	59.0
1¼	32	2.89	4.30	69.4	63.0	79.9	72.5
1⅜	35	3.50	5.21	83.5	75.7	96.0	87.1
1½	38	4.16	6.19	98.9	89.7	114	103
1⅝	42	4.88	7.26	115	104	132	120
1¾	45	5.67	8.44	133	121	153	139
1⅞	48	6.5	9.67	152	138	174	158
2	51	7.39	11.0	172	156	198	180
2⅛	54	8.35	12.4	192	174	221	200
2¼	57	9.36	13.9	215	195	247	224
2⅜	61	10.4	15.5	239	217	274	249
2½	64	11.6	17.3	262	238	302	274
2⅝	67	12.8	19.0	288	261	331	300
2¾	70	14.0	20.8	314	285	361	327
2⅞	74	15.3	22.8	341	309	392	356
3	77	16.6	24.7	370	336	425	386
3⅛	80	18.0	26.8	399	362	458	415
3¼	83	19.5	29.0	429	389	492	446
3⅜	86	21.0	31.3	459	416	529	480
3½	90	22.7	33.8	491	445	564	512
3⅝	96	24.3	36.2	523	458	602	528
3¾	103	26.0	38.7	557	505	641	581

* To convert to kilonewtons (kN), multiply tons (nominal breaking strength) by 8.896; 1 lb = 4.448 N.

Table A.10 Nominal Strengths of Wire Rope
6 × 61 Classification/bright (uncoated), fiber core

Nominal Diameter		Approximate Mass		Nominal Strength* Improved Plow Steel	
in	mm	lb/ft	kg/m	tons	t
1	26	1.68	2.5	39.8	36.1
1⅛	29	2.13	3.17	50.1	45.4
1¼	32	2.63	3.91	61.5	55.8
1⅜	35	3.18	4.73	74.1	67.2
1½	38	3.78	5.63	87.9	79.7
1⅝	42	4.44	6.61	103	93.4
1¾	45	5.15	7.66	119	108
1⅞	48	5.91	8.80	136	123
2	51	6.77	10.1	154	140
2⅛	54	7.59	11.3	173	157
2¼	57	8.51	12.7	193	175
2⅜	61	9.48	14.1	214	194
2½	64	10.5	15.6	236	214
2⅝	67	11.6	17.3	260	236
2¾	70	12.7	18.9	284	258
2⅞	74	13.9	20.7	309	280
3	77	15.1	22.5	335	304
3¼	83	17.7	26.3	390	354
3⅜	86	19.1	28.4	419	380
3½	90	20.6	30.7	449	407
3¾	96	23.6	35.1	511	464
4	103	26.9	40.0	577	523
4¼	109	30.3	45.1	646	586
4½	115	34.0	50.6	719	652
4¾	122	37.9	56.4	794	720
5	128	42.0	62.5	872	791

* To convert to kilonewtons (kN), multiply tons (nominal breaking strength) by 8.896; 1 lb = 4.448 N.

Table A.11 Nominal Strengths of Wire Rope
6 × 61 Classification/bright (uncoated), IWRC

Nominal Diameter		Approximate Mass		Nominal Strength* Improved Plow Steel		Extra Improved Plow Steel	
in	mm	lb/ft	kg/m	tons	t	tons	t
1	26	1.85	2.75	42.8	38.8	49.1	44.5
1⅛	29	2.34	3.48	53.9	48.9	61.9	56.2
1¼	32	2.89	4.30	66.1	60.0	76.1	69.0
1⅜	35	3.50	6.59	79.7	72.3	91.7	83.2
1½	38	4.16	6.19	94.5	85.7	109	98.9
1⅝	42	4.88	7.62	111	101	127	115
1¾	45	5.67	8.44	128	116	146	132
1⅞	48	6.50	9.67	146	132	168	152
2	51	7.39	11.0	165	150	190	172
2⅛	54	8.35	12.4	186	169	214	194
2¼	57	9.36	13.9	207	188	239	217
2⅜	61	10.40	15.5	230	209	264	240
2½	64	11.6	17.3	254	230	292	265
2⅝	67	12.8	18.3	279	253	321	291
2¾	70	14.0	20.8	305	277	350	318
2⅞	74	15.3	22.8	333	302	382	347
3	77	16.6	24.7	360	327	414	376
3¼	83	19.5	29.0	419	380	483	438
3⅜	86	21.0	31.3	451	409	518	470
3½	90	22.7	33.8	483	438	555	503
3¾	96	26.0	38.7	549	498	632	573
4	103	29.6	44.1	620	562	713	647
4¼	109	33.3	49.6	694	630	799	725
4½	115	37.4	55.7	772	700	888	806
4¾	122	41.7	62.1	853	774	981	890
5	128	46.2	68.8	937	850	1078	978

* To convert to kilonewtons (kN), multiply tons (nominal breaking strength) by 8.896; 1 lb = 4.448 N.

Table A.12 Nominal Strengths of Wire Rope
6 × 91 Classification/bright (uncoated), fiber core

Nominal Diameter		Approximate Mass		Nominal Strength* Improved Plow Steel	
in	mm	lb/ft	kg/m	tons	t
2	51	6.77	10.1	146	132
2⅛	54	7.59	11.3	164	149
2¼	57	8.51	12.7	183	166
2⅜	61	9.48	14.1	203	184
2½	64	10.5	15.6	225	204
2⅝	67	11.6	17.3	247	224
2¾	70	12.7	18.9	270	245
3	77	15.1	22.5	318	288
3¼	83	17.7	26.3	371	337
3½	90	20.6	30.7	426	386

* To convert to kilonewtons (kN), multiply tons (nominal breaking strength) by 8.896; 1 lb = 4.448 N.

Table A.13 Nominal Strengths of Wire Rope
6 × 91 Classification/bright (uncoated), IWRC

Nominal Diameter		Approximate Mass		Nominal Strength*			
				Improved Plow Steel		Extra Improved Plow Steel	
in	mm	lb/ft	kg/m	tons	t	tons	t
2	51	7.39	11.0	157	142	181	164
2⅛	54	8.35	12.4	176	160	203	184
2¼	57	9.36	14.0	197	179	227	206
2⅜	61	10.4	15.5	218	198	251	228
2½	64	11.6	17.3	242	220	277	251
2⅝	67	12.8	19.0	265	240	305	277
2¾	70	14.0	20.8	290	263	333	302
3	77	16.6	24.7	342	310	393	357
3¼	83	19.5	29.0	399	362	458	415
3½	90	22.7	33.8	458	415	527	478
3¾	96	26.0	38.7	522	474	600	544
4	103	29.6	44.1	589	534	677	614
4¼	109	33.3	49.6	660	599	759	689
4½	115	37.4	55.7	734	666	844	766
4¾	122	41.7	62.1	810	787	932	846
5	128	46.2	68.7	891	808	1024	929
5¼	134	49.8	74.1	974	884	1120	1016
5½	140	54.5	81.1	1060	962	1219	1106
5¾	146	59.6	88.7	1148	1041	1320	1198
6	153	65.0	96.7	1240	1125	1426	1294

* To convert kilonewtons (kN), multiply tons (nominal breaking strength) by 8.896; 1 lb = 4.448 N.

Table A.14 Nominal Strengths of Wire Rope
6 × 30, 6 × 30G, 6 × 25B, 6 × 27H Flattened strand/fiber core

Nominal Diameter		Approximate Mass		Nominal Strength* Improved Plow Steel	
in	mm	lb/ft	kg/m	tons	t
½	13	0.45	0.67	11.8	10.8
9⁄16	14.5	0.57	0.85	14.9	13.5
⅝	16	0.70	1.04	18.3	16.6
¾	19	1.01	1.50	26.2	23.8
⅞	22	1.39	2.07	35.4	32.1
1	26	1.80	2.68	46.0	41.7
1⅛	29	2.28	3.39	57.9	52.5
1¼	32	2.81	4.18	71.0	64.4
1⅜	35	3.40	5.06	85.5	77.6
1½	38	4.05	6.03	101	91.6
1⅝	42	4.75	7.07	118	107
1¾	45	5.51	8.20	136	123
1⅞	48	6.33	9.42	155	141
2	51	7.20	10.70	176	160

* To convert to kilonewtons (kN), multiply tons (nominal breaking strength) by 8.896; 1 lb = 4.448 N.

Table A.15 Nominal Strengths of Wire Rope
6 × 30, 6 × 30G, 6 × 25B, 6 × 27H Flattened strand/IWRC*

Nominal Diameter		Approximate Mass		Nominal Strength*			
				Improved Plow Steel		Extra Improved Plow Steel	
in	mm	lb/ft	kg/m	tons	t	tons	t
½	13	0.47	0.70	12.6	11.4	14	12.7
9⁄16	14.5	0.60	0.89	16.0	14.5	17.6	16
⅝	16	0.73	1.09	19.6	17.8	21.7	19.7
¾	19	1.06	1.58	28.1	25.5	31	28.1
⅞	22	1.46	2.17	38.0	34.5	41.9	38
1	26	1.89	2.83	49.4	44.8	54.4	49.4
1⅛	29	2.39	3.56	62.2	56.4	68.5	62.1
1¼	32	2.95	4.39	76.3	69.2	84	76.2
1⅜	35	3.57	5.31	91.9	83.4	101	91.6
1½	38	4.25	6.32	108	98	119	108
1⅝	42	4.98	7.41	127	115	140	127
1¾	45	5.78	8.60	146	132	161	146
1⅞	48	6.65	9.90	167	152	184	167
2	51	7.56	11.3	189	171	207	188

* To convert to kilonewtons (kN), multiply tons (nominal breaking strength) by 8.896; 1 lb = 4.448 N.

Table A.16 Nominal Strengths of Wire Rope
8 × 19 Classification/bright (uncoated), fiber core

Nominal Diameter		Approximate Mass		Nominal Strength* Improved Plow Steel	
in	mm	lb/ft	kg/m	tons	t
¼	6.5	0.10	0.15	2.35	2.13
⁵⁄₁₆	8	0.15	0.22	3.65	3.31
⅜	9.5	0.22	0.33	5.24	4.75
⁷⁄₁₆	11.5	0.30	0.45	7.09	6.43
½	13	0.39	0.58	9.23	8.37
⁹⁄₁₆	14.5	0.50	0.74	11.6	10.5
⅝	16	0.61	0.91	14.3	13.0
¾	19	0.88	1.31	20.5	18.6
⅞	22	1.20	1.79	27.7	25.1
1	26	1.57	2.34	36.0	32.7
1⅛	29	1.99	2.96	45.3	41.1
1¼	32	2.45	3.65	55.7	50.5
1⅜	35	2.97	4.42	67.1	60.7
1½	38	3.53	5.25	79.4	72.0

* To convert to kilonewtons (kN), multiply tons (nominal breaking strength) by 8.896; 1 lb = 4.448 N.

Table A.17 Nominal Strengths of Wire Rope
8 × 19 Classification/bright (uncoated), IWRC

Nominal Diameter		Approximate Mass		Nominal Strength*			
				Improved Plow Steel		Extra Improved Plow Steel	
in	mm	lb/ft	kg/m	tons	t	tons	t
½	13	0.47	0.70	10.1	9.16	11.6	10.5
⁹⁄₁₆	14.5	0.60	0.89	12.8	11.6	14.7	13.3
⅝	16	0.73	1.09	15.7	14.2	18.1	16.4
¾	19	1.06	1.58	22.5	20.4	25.9	23.5
⅞	22	1.44	2.14	30.5	27.7	35.0	31.8
1	26	1.88	2.80	39.6	35.9	45.5	41.3
1⅛	29	2.39	3.56	49.8	45.2	57.3	51.7

* To convert to kilonewtons (kN), multiply tons (nominal breaking strength) by 8.896; 1 lb = 4.448 N.

Table A.18 Nominal Strengths of Wire Rope
18 × 7 Construction/bright (uncoated) or drawn-galvanized wire, fiber core

Nominal Diameter		Approximate Mass		Nominal Strength*			
				Improved Plow Steel		Extra Improved Plow Steel	
in	mm	lb/ft	kg/m	tons	t	tons	t
½	13	0.43	0.64	9.85	8.94	10.8	9.8
⁹⁄₁₆	14.5	0.55	0.82	12.4	11.2	13.6	12.3
⅝	16	0.68	1.01	15.3	18.9	16.8	15.2
¾	19	0.97	1.44	21.8	19.8	24.0	21.8
⅞	22	1.32	1.96	29.5	26.8	32.5	29.5
1	26	1.73	2.57	38.3	34.7	42.2	38.3
1⅛	29	2.19	3.26	48.2	43.7	53.1	48.2
1¼	32	2.70	4.02	59.2	53.7	65.1	59.1
1⅜	35	3.27	4.87	71.3	64.7	78.4	1.1
1½	38	3.89	5.79	84.4	76.6	92.8	84.2

* To convert to kilonewtons (kN), multiply tons (nominal breaking strength) by 8.896; 1 lb = 4.448 N.

Table A.19 Nominal Strengths of Wire Rope
19 × 7 Construction/bright (uncoated) or drawn-galvanized wire

Nominal Diameter		Approximate Mass		Nominal Strength*			
				Improved Plow Steel		Extra Improved Plow Steel	
in	mm	lb/ft	kg/m	tons	t	tons	t
½	13	0.45	0.67	9.85	8.94	10.8	9.8
⁹⁄₁₆	14.5	0.58	0.86	12.4	11.2	13.6	12.3
⅝	16	0.71	1.06	15.3	13.9	16.8	15.2
¾	19	1.02	1.52	21.8	19.8	24.0	21.8
⅞	22	1.39	2.07	29.5	26.8	32.5	29.5
1	26	1.82	2.71	38.3	34.7	42.2	38.3
1⅛	29	2.30	3.42	48.2	43.7	53.1	48.2
1¼	32	2.84	4.23	59.2	53.7	65.1	59.1
1⅜	35	3.43	5.10	71.3	64.7	78.4	71.1
1½	38	4.08	6.07	84.4	76.6	92.8	84.2

* To convert to kilonewtons (kN), multiply tons (nominal breaking strength) by 8.896; 1 lb = 4.448 N.

Table A.20 Nominal Strengths of Wire Rope
6 × 7 Classification/galvanized, fiber core*

Nominal Diameter		Approximate Mass		Nominal Strength† Improved Plow Steel	
in	mm	lb/ft	kg/m	tons	t
¼	6.5	0.09	0.14	2.38	2.16
⁵⁄₁₆	8	0.15	0.22	3.69	3.35
⅜	9.5	0.21	0.31	5.27	4.78
⁷⁄₁₆	11.5	0.29	0.43	7.14	6.48
½	13	0.38	0.57	9.27	8.41
⁹⁄₁₆	14.5	0.48	0.71	11.7	10.6
⅝	16	0.59	0.88	14.3	13.0
¾	19	0.84	1.25	20.4	18.5
⅞	22	1.15	1.71	27.6	25.0
1	26	1.50	2.23	35.7	32.4
1⅛	29	1.90	2.83	44.8	40.6
1¼	32	2.34	3.48	54.9	49.8
1⅜	35	2.84	4.23	65.8	59.7
1½	38	3.38	5.03	77.6	70.4

* For ropes with an IWRC, add 7½ percent to their respective *nominal strengths* and 10 percent to their *approximate mass* (weights). Fiber cores consist either of polypropylene or natural fiber.

† To convert to kilonewtons (kN), multiply tons (nominal breaking strength) by 8.896; 1 lb = 4.448 N.

Table A.21 Nominal Strengths of Wire Rope
6 × 12 Construction/galvanized, fiber core*

Nominal Diameter		Approximate Mass		Nominal Strength† Improved Plow Steel	
in	mm	lb/ft	kg/m	tons	t
⁵⁄₁₆	8	0.10	0.14	2.34	2.12
⅜	9.5	0.15	0.22	3.36	3.05
⁷⁄₁₆	11.5	0.20	0.30	4.55	4.13
½	13	0.26	0.39	5.91	4.71
⁹⁄₁₆	14.5	0.33	0.49	7.45	6.76
⅝	16	0.41	0.61	9.16	8.31
¾	19	0.59	0.88	13.1	11.9
¹³⁄₁₆	21	0.69	1.03	15.3	13.9
⅞	22	0.80	1.19	17.7	16.1
1	26	1.05	1.56	23.0	20.9

* Fiber cores consist either of polypropylene or natural fiber.
† To convert to kilonewtons (kN), multiply tons (nominal breaking strength) by 8.896; 1 lb = 4.448 N.

Table A.22 Nominal Strengths of Wire Rope
6 × 24 Construction/galvanized, fiber core*

Nominal Diameter		Approximate Mass		Nominal Strength† Improved Plow Steel	
in	mm	lb/ft	kg/m	tons	t
⅜	9.5	0.19	0.29	4.77	4.33
½	13	0.35	0.52	8.40	7.62
⁹⁄₁₆	14.5	0.44	0.65	10.6	9.62
⅝	16	0.54	0.80	13.0	11.8
¾	19	0.78	1.16	18.6	16.9
⅞	22	1.06	1.58	25.2	22.9
1	26	1.38	2.05	32.8	29.8
1⅛	29	1.75	2.60	41.2	37.4
1¼	32	2.16	3.21	50.7	46.0
1⅜	35	2.61	3.88	61.0	55.3
1½	38	3.11	4.63	72.3	65.6
1⅝	42	3.64	5.42	84.5	76.7
1¾	45	4.23	6.30	97.5	88.5
1⅞	48	4.85	7.22	111	101
2	51	5.52	8.21	126	114

* Fiber cores consist either of polypropylene or natural fiber.
† To convert to kilonewtons (kN), multiply tons (nominal breaking strength) by 8.896; 1 lb = 4.448 N.

WIRE ROPE ASSEMBLIES

When ordering *wire rope with fittings attached,* lengths—as shown in Figure A.9—should be specified. Additionally, the load at which this measurement is taken should be specified, i.e., at no load, at a percentage of catalog breaking strength, etc.

The accompanying drawings do not show all possible combinations of fittings; in any case, the same measuring methods should be followed.

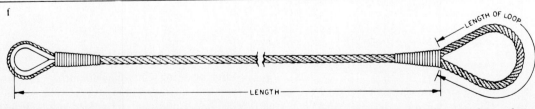

FIG. A.9 Wire rope assemblies. *(a)* Zinc-attached closed wire rope socket at one end; zinc-attached open wire rope socket at other end. Measurement: Pull of closed socket to centerline of open socket pin. *(b)* Closed swaged wire rope socket at one end; open swaged wire rope socket at the other end. Measurement: Centerline of pin to centerline of pin. *(c)* Closed bridge socket attached to one end; open bridge socket attached to other end. Measurements: Centerline of closed socket pin to centerline of open socket pin; include two of the three values: take-up, contraction, and expansion. The values of C and O are also required. *(d)* Thimble spliced at one end. Measurement: Pull of thimble to end of rope. *(e)* Link spliced at one end; hook spliced at other end. Measurement: Pull of link to pull of hook. *(f)* Thimble spliced at one end; loop spliced at other end. Measurements: Pull of thimble to base of loop and circumference of loop.

B

Rivets and Threaded Fasteners*

* This appendix is adapted from Manual of Steel
Construction, American Institute of Steel Construction,
New York, 1970.

Table B.1 Tension
Allowable loads in kips

Unfinished Bolts and Threaded Parts
Tension on Tensile Stress Area

ASTM Designation or Yield Stress	Allowable Tensile Stress F_g ksi	Nominal Diameter, in							
		⅝	¾	⅞	1	1⅛	1¼	1⅜	1½
		Tensile Stress Area,* in²							
		0.2260	0.3345	0.4617	0.6057	0.7633	0.9691	1.1549	1.4053
A307 Bolts	20.0	4.52	6.69	9.23	12.11	15.27	19.38	23.10	28.11
Threaded parts F_y, ksi									
36	22.0	4.97	7.36	10.16	13.33	16.79	21.32	25.41	30.92
42	25.2	5.70	8.41	11.64	15.27	19.23	24.42	29.23	35.53
45	27.0	6.10	9.03	12.47	16.35	20.61	26.17	31.18	37.94
50	30.0	6.78	10.04	13.85	18.17	22.90	29.07	34.65	42.16
55	33.0	7.46	11.04	15.24	19.99	25.19	31.98	38.11	46.37
60	36.0	8.14	12.04	16.62	21.81	—	—	—	—

* The definition of tensile stress area is given in the AISC Specification, Section 1.5.2.1. Values are based on UNC thread dimensions.

Nuts must meet specifications compatible with threaded parts.

For upset rods see AISC Specification, Section 1.5.2.1.

Rivets and High Strength Bolts
Tension on Gross (Nominal) Area

ASTM Designation	Allowable Tensile Stress F_t, ksi	Nominal Diameter, in							
		⅝	¾	⅞	1	1⅛	1¼	1⅜	1½
		Gross (Nominal) Area, in²							
		0.3068	0.4418	0.6013	0.7854	0.9940	1.2272	1.4849	1.7671
Rivets									
A502-1	20.0	6.14	8.84	12.03	15.71	19.88	24.54	29.70	35.34
A502-2	27.0	8.28	11.93	16.24	21.21	26.84	33.13	40.09	47.71
Bolts									
A325	40.0	12.27	17.67	24.05	31.42	39.76	49.09	59.40	70.68
A490	54.0†	16.57†	23.86†	32.47†	42.41†	53.68†	66.27†	80.18†	95.42†

† For static loading only.

For allowable combined shear and tension loads, see AISC Specification, Section 1.6.3.

Table B.2 Shear
Allowable loads in kips

	Power-Driven Shop and Field Rivets			
Diameter—Area	$^5/_8$ in — 0.3068 in^2		$^3/_4$ in — 0.4418 in^2	
ASTM designation	A502-1	A502-2	A502-1	A502-2
Shear F_v, ksi	15.0	20.0	15.0	20.0
Single shear, kips	4.60	6.14	6.63	8.84
Double shear, kips	9.20	12.27	13.25	17.67

	Unfinished Bolts, ASTM A307, and Threaded Parts, $F_y = 36$ ksi*			
Diameter—Area	$^5/_8$ in — 0.3068 in^2		$^3/_4$ in — 0.4418 in^2	
ASTM designation or yield stress, ksi	A307	$F_y = 36$	A307	$F_y = 36$
Shear F_v, ksi	10.0	10.8*	10.0	10.8*
Single shear, kips	3.07	3.31	4.42	4.77
Double shear, kips	6.14	6.63	8.84	9.54

	High-Strength Bolts in Friction-Type Connections and in Bearing-Type Connections with Threads in Shear Planes					
Diameter—Area	$^5/_8$ in — 0.3068 in^2			$^3/_4$ in — 0.4418 in^2		
ASTM designation†	A325-F A325-N	A490-F	A490-N	A325-F A325-N	A490-F	A490-N
Shear F_v, ksi	15.0	20.0	22.5	15.0	20.0	22.5
Single shear, kips	4.60	6.14	6.90	6.63	8.84	9.94
Double shear, kips	9.20	12.27	13.81	13.25	17.67	19.88

	High-Strength Bolts in Bearing-Type Connections with Threads Excluded from Shear Planes			
Diameter—Area	$^5/_8$ in — 0.3068 in^2		$^3/_4$ in — 0.4418 in^2	
ASTM designation†	A325-X	A490-X	A325-X	A490-X
Shear F_v, ksi	22.0	32.0	22.0	32.0
Single shear, kips	6.75	9.82	9.72	14.14
Double shear, kips	13.50	19.64	19.44	28.28

* For threaded parts of material other than $F_y = 36$ ksi steel, use $F_v = 0.30\,F_y$.
† The letter suffixes following the ASTM designations A325 and A490 represent the following:
  F: Friction-type connection
  N: Bearing-type connection with threads included in shear plane
  X: Bearing-type connection with threads excluded from shear plane

Table B.3 Bearing
Allowable loads in kips
All rivets and bolts in bearing-type connections

Diameter, in	⁵⁄₈								³⁄₄							
F_y, ksi	36	42	45	50	55	60	65	100	36	42	45	50	55	60	65	100
Bearing F_p, ksi	48.6	56.7	60.8	67.5	74.3	81.0	87.8	135	48.6	56.7	60.8	67.5	74.3	81.0	87.8	135
Material Thickness																
⅛	3.80	4.43	4.75	5.27	5.80	6.33	6.86	10.5	4.56	5.31	5.70	6.33	6.97	7.59	8.23	12.7
³⁄₁₆	5.70	6.64	7.13	7.91	8.71	9.49	10.3	15.8	6.83	7.97	8.55	9.49	10.4	11.4	12.3	19.0
¼	7.59	8.86	9.50	10.6	11.6	12.7	13.7	21.1	9.11	10.6	11.4	12.7	13.9	15.2	16.5	25.3
⁵⁄₁₆	9.49	11.1	11.9	13.2	14.5	15.8	17.1	26.4	11.4	13.3	14.3	15.8	17.4	19.0	20.6	31.6
⅜	11.4	13.3	14.3	15.8	17.4	19.0	20.6	31.6	13.7	15.9	17.1	19.0	20.9	22.8	24.7	38.0
⁷⁄₁₆	13.3	15.5	16.6	18.5	20.3	22.1		36.9	15.9	18.6	19.9	22.1	24.4	26.6	28.8	44.3
½	15.2	17.7	19.0	21.1				42.2	18.2	21.3	22.8	25.3	27.9	30.4		50.6
⁹⁄₁₆	17.1	19.9	21.4					47.5	20.5	23.9	25.7	28.5	31.3			57.0
⅝	19.0							52.7	22.8	26.6	28.5					63.3
¹¹⁄₁₆	20.9							58.0	25.1	29.2						69.6
¾								63.3	27.3							75.9
¹³⁄₁₆								68.6	29.6							82.3
⅞								73.8								88.6
¹⁵⁄₁₆								79.1								94.9
1	30.4	35.4	38.0	42.2	46.4	50.6	54.9	84.4	36.5	42.5	45.6	50.6	55.7	60.8	65.9	101

Note: This table not applicable to fasteners in friction-type connections.

F_y is the yield stress of the connected material; see AISC specification, Section 1.5.2.2.

F_p, the unit bearing stress, applies equally to conditions of single shear and enclosed bearing.

Values for thicknesses not listed may be obtained by multiplying the unlisted thickness by the value given for a 1-in thickness in the appropriate F_y column.

Values for F_y not listed may be obtained by multiplying the value given for $F_y = 100$ ksi by the unlisted F_y and dividing by 100.

Table B.4   Shear
Allowable loads in kips

Power-Driven Shop and Field Rivets				
Diameter — Area	⁷/₈ in — 0.6013 in²		1 in — 0.7854 in²	
ASTM designation	A502-1	A502-2	A502-1	A502-2
Shear F_v, ksi	15.0	20.0	15.0	20.0
Single shear, kips Double shear, kips	9.02 18.04	12.03 24.05	11.78 23.56	15.71 31.42

Unfinished Bolts, ASTM A307, and Threaded Parts, $F_y = 36$ ksi*				
Diameter — Area	⁷/₈ in — 0.6013 in²		1 in — 0.7854 in²	
ASTM designation or yield stress, ksi	A307	$F_y = 36$	A307	$F_y = 36$
Shear F_v, ksi	10.0	10.8*	10.0	10.8*
Single shear, kips Double shear, kips	6.01 12.03	6.49 12.99	7.85 15.71	8.48 16.96

High-Strength Bolts in Friction-Type Connections and in Bearing-Type Connections with Threads in Shear Planes						
Diameter — Area	⁷/₈ in — 0.6013 in²			1 in — 0.7854 in²		
ASTM designation†	A325-F A325-N	A490-F	A490-N	A325-F A325-N	A490-F	A490-N
Shear F_v, ksi	15.0	20.0	22.5	15.0	20.0	22.5
Single shear, kips Double shear, kips	9.02 18.04	12.03 24.05	13.53 27.06	11.78 23.56	15.71 31.42	17.67 35.34

High-Strength Bolts in Bearing-Type Connections with Threads Excluded from Shear Planes				
Diameter — Area	⁷/₈ in — 0.6013 in²		1 in — 0.7854 in²	
ASTM designation†	A325-X	A490-X	A325-X	A490-X
Shear F_v, ksi	22.0	32.0	22.0	32.0
Single shear, kips Double shear, kips	13.23 26.46	19.24 38.48	17.28 34.56	25.13 50.27

* For threaded parts of material other than $F_y = 36$ ksi steel, use $F_v = 0.30\, F_y$.
† The letter suffixes following the ASTM designations A325 and A490 represent the following:
F: Friction-type connection
N: Bearing-type connection with threads included in shear plane
X: Bearing-type connection with threads excluded from shear plane

Table B.5   Bearing
Allowable loads in kips
All rivets and bolts in bearing-type connections

Diameter, in	⁷⁄₈								1							
F_y, ksi	36	42	45	50	55	60	65	100	36	42	45	50	55	60	65	100
Bearing F_p, ksi	48.6	58.7	60.8	67.5	74.3	81.0	87.8	135	48.6	56.7	60.8	67.5	74.3	81.0	87.8	135
Material Thickness																
⅛	5.32	6.20	6.65	7.38	8.13	8.86	9.60	14.8	6.08	7.09	7.60	8.44	9.29	10.1	11.0	16.9
³⁄₁₆	7.97	9.30	9.98	11.1	12.2	13.3	14.4	22.1	9.11	10.6	11.4	12.7	13.9	15.2	16.5	25.3
¼	10.6	12.4	13.3	14.8	16.3	17.7	19.2	29.5	12.2	14.2	15.2	16.9	18.6	20.3	22.0	33.8
⁵⁄₁₆	13.3	15.5	16.6	18.5	20.3	22.1	24.0	36.9	15.2	17.7	19.0	21.1	23.2	25.3	27.4	42.2
⅜	15.9	18.6	20.0	22.1	24.4	26.6	28.8	44.3	18.2	21.3	22.8	25.3	27.9	30.4	32.9	50.6
⁷⁄₁₆	18.6	21.7	23.3	25.8	28.4	31.0	33.6	51.7	21.3	24.8	26.6	29.5	32.5	35.4	38.4	59.1
½	21.3	24.8	26.6	29.5	32.5	35.4	38.4	59.1	24.3	28.4	30.4	33.8	37.2	40.5	43.9	67.5
⁹⁄₁₆	23.9	27.9	29.9	33.2	36.6	39.9	43.2	66.4	27.3	31.9	34.2	38.0	41.8	45.6	49.4	75.9
⅝	26.6	31.0	33.3	36.9	40.6			73.8	30.4	35.4	38.0	42.2	46.4	50.6	54.9	84.4
¹¹⁄₁₆	29.2	34.1	36.6	40.6				81.2	33.4	39.0	41.8	46.4	51.1			92.8
¾	31.9	37.2	39.9					88.6	36.5	42.5	45.6	50.6				101
¹³⁄₁₆	34.6	40.3						96.0	39.5	46.1	49.4					110
⅞	37.2							103	42.5	49.6	53.2					118
¹⁵⁄₁₆	39.9							111	45.6	53.2						127
1	42.5	49.6	53.2	59.1	65.0	70.9	76.8	118	48.6	56.7	60.8	67.5	74.3	81.0	87.8	135
1¹⁄₁₆								126	51.6							143

Note: This table not applicable to fasteners in friction-type connections.

F_y is the yield stress of the connected material; see AISC specification, Section 1.5.2.2.

F_p, the unit bearing stress, applies equally to conditions of single shear and enclosed bearing.

Values for thicknesses not listed may be obtained by multiplying the unlisted thickness by the value given for a 1-in thickness in the appropriate F_y column.

Values for F_y not listed may be obtained by multiplying the value given for $F_y = 100$ ksi by the unlisted F_y and dividing by 100.

Table B.6 Shear
Allowable loads in kips

Power-Driven Shop and Field Rivets				
Diameter—Area	1⅛ in—0.9940 in²		1¼ in—1.2272 in²	
ASTM designation	A502-1	A502-2	A502-1	A502-2
Shear F_v, ksi	15.0	20.0	15.0	20.0
Single shear, kips	14.91	19.88	18.41	24.54
Double shear, kips	29.82	39.76	36.82	49.09

Unfinished Bolts, ASTM A307, and Threaded Parts, F_y = 36 ksi*				
Diameter—Area	1⅛ in—0.9940 in²		1¼ in—1.2272 in²	
ASTM designation or yield stress, ksi	A307	F_y = 36	A307	F_y = 36
Shear F_v, ksi	10.0	10.8*	10.0	10.8*
Single shear, kips	9.94	10.74	12.27	13.25
Double shear, kips	19.88	21.47	24.54	26.51

High-Strength Bolts in Friction-Type Connections and in Bearing-Type Connections with Threads in Shear Planes					
Diameter—Area	1⅛ in—0.9940 in²			1¼ in—1.2272 in²		
ASTM designation†	A325-F A325-N	A490-F	A490-N	A325-F A325-N	A490-F	A490-N
Shear F_v, ksi	15.0	20.0	22.5	15.0	20.0	22.5
Single shear, kips	14.91	19.88	22.37	18.41	24.54	27.61
Double shear, kips	29.82	36.76	44.73	36.82	49.09	55.22

High-Strength Bolts in Bearing-Type Connections with Threads Excluded from Shear Planes				
Diameter—Area	1⅛ in—0.9940 in²		1¼ in—1.2272 in²	
ASTM designation†	A325-X	A490-X	A325-X	A490-X
Shear F_v, ksi	22.0	32.0	22.0	32.0
Single shear, kips	21.87	31.81	27.00	39.27
Double shear, kips	43.74	63.62	54.00	78.54

* For threaded parts of material other than F_y = 36 ksi steel, use F_v = 0.30 F_y.
† The letter suffixes following the ASTM designations A325 and A490 represent the following:
F: Friction-type connection
N: Bearing-type connection with threads included in shear plane
X: Bearing-type connection with threads excluded from shear plane

Table B.7 Bearing
Allowable loads in kips
All rivets and bolts in bearing-type connection

Diameter, in	1⅛								1¼							
F_y, ksi	36	42	45	50	55	60	65	100	36	42	45	50	55	60	65	100
Bearing F_p, ksi	48.6	56.7	60.8	67.5	74.3	81.0	87.8	135	48.6	56.7	60.8	67.5	74.3	81.0	87.8	135
Material Thickness																
⅛	6.83	7.97	8.55	9.49	10.4	11.4	12.3	19.0	7.59	8.86	9.50	10.6	11.6	12.7	13.7	21.1
3/16	10.2	12.0	12.8	14.2	15.7	17.1	18.5	28.5	11.4	13.3	14.3	15.8	17.4	19.0	20.6	31.6
¼	13.7	15.9	17.1	19.0	20.9	22.8	24.7	38.0	15.2	17.7	19.0	21.1	23.2	25.3	27.4	42.2
5/16	17.1	19.9	21.4	23.7	26.1	28.5	30.9	47.5	19.0	22.1	23.7	26.4	29.0	31.6	34.3	52.7
⅜	20.5	23.9	25.7	28.5	31.3	34.2	37.0	57.0	22.8	26.6	28.5	31.6	34.8	38.0	41.2	63.3
7/16	23.9	27.9	29.9	33.2	36.6	39.9	43.2	66.4	26.6	31.0	33.3	36.9	40.6	44.3	48.0	73.8
½	27.3	31.9	34.2	38.0	41.8	45.6	49.4	75.9	30.4	35.4	38.0	42.2	46.4	50.6	54.9	84.4
9/16	30.8	35.9	38.5	42.7	47.0	51.3	55.6	85.4	34.2	39.9	42.7	47.5	52.2	57.0	61.7	94.9
⅝	34.2	39.9	42.7	47.5	52.2	57.0	61.7	94.9	38.0	44.3	47.5	52.7	58.1	63.3	68.6	105
11/16	37.6	43.9	47.0	52.2	57.5	62.6	67.9	104	41.8	48.7	52.5	58.0	63.9	69.6	75.5	116
¾	41.0	47.8	51.3	57.0	62.7	68.3		114	45.6	53.2	57.0	63.3	69.7	75.9	82.3	127
13/16	44.4	51.8	55.6	61.7	67.9			123	49.4	57.6	61.7	68.6	75.5	82.3		137
⅞	47.8	55.8	59.9	66.4				133	53.2	62.0	66.5	73.8	81.3			148
15/16	51.3	59.8	64.1					142	57.0	66.4	71.3	79.1				158
1	54.7	63.8	68.4	75.9	83.6	91.1	98.8	152	60.8	70.9	76.0	84.4	92.9	101	110	169
1 1/16	58.1							161	64.5	75.3	80.7					179
1⅛	61.5							171	68.3	79.7						190
1 3/16	64.9							180	72.1							200
1¼									75.9							211
1 5/16									79.7							221

Note: This table not applicable to fasteners in friction-type connections.
F_y is the yield stress of the connected material; see AISC specification, Section 1.5.2.2.
F_p, the unit bearing stress, applies equally to conditions of single shear and enclosed bearing.
Values for thicknesses not listed may be obtained by multiplying the unlisted thickness by the value given for a 1-in thickness in the appropriate F_y column.
Values for F_y not listed may be obtained by multiplying the value given for $F_y = 100$ ksi by the unlisted F_y and dividing by 100.

Table B.8 Shear
Allowable load in kips

Power Driven Shop and Field Rivets

Diameter — Area	1³/₈ in — 1.4849 in²		1¹/₂ in — 1.7671 in²	
ASTM designation	A502-1	A502-2	A502-1	A502-2
Shear F_v, ksi	15.0	20.0	15.0	20.0
Single shear, kips Double shear, kips	22.27 44.55	29.70 59.40	26.51 53.01	35.34 70.68

Unfinished Bolts, ASTM A307, and Threaded Parts, $F_y = 36$ ksi*

Diameter — Area	1³/₈ in — 1.4849 in²		1¹/₂ in — 1.7621 in²	
ASTM designation or yield stress, ksi	A307	$F_y = 36$	A307	$F_y = 36$
Shear F_v, ksi	10.0	10.8*	10.0	10.8*
Single shear, kips Double shear, kips	14.85 29.70	16.04 32.07	17.67 35.34	19.08 38.17

High-Strength Bolts in Friction-Type Connections and in Bearing-Type Connections with Threads in Shear Planes

Diameter — Area	1³/₈ in — 1.4849 in²			1¹/₂ in — 1.7671 in²		
ASTM designation†	A325-F A325-N	A490-F	A490-N	A325-F A325-N	A490-F	A490-N
Shear F_v, ksi	15.0	20.0	22.5	15.0	20.0	22.5
Single shear, kips Double shear, kips	22.27 44.55	29.70 59.40	33.41 66.82	26.51 53.01	35.34 70.68	39.76 79.52

High-Strength Bolts in Bearing-Type Connections with Threads Excluded from Shear Planes

Diameter — Area	1³/₈ in — 1.4849 in²		1¹/₂ in — 1.7671 in²	
ASTM designation†	A325-X	A490-X	A325-X	A490-X
Shear F_v, ksi	22.0	32.0	22.0	32.0
Single shear, kips Double shear, kips	32.67 65.34	47.52 95.03	38.88 77.75	56.55 113.09

* For threaded parts of material other than $F_y = 36$ ksi steel, use $F_v = 0.30 F_y$.
† The letter suffixes following the ASTM designations A325 and A490 represent the following:
F: Friction-type connection
N: Bearing-type connection with threads included in shear plane
X: Bearing-type connection with threads excluded from shear plane

Table B.9 Bearing
Allowable loads in kips
All rivets and bolts in bearing-type connections

Diameter, in	$1\frac{3}{8}$								$1\frac{1}{2}$							
F_y, ksi	36	42	45	50	55	60	65	100	36	42	45	50	55	60	65	100
Bearing F_p, ksi	48.6	56.7	60.8	67.5	74.3	81.0	87.8	135	48.6	56.7	60.8	67.5	74.3	81.0	87.8	135
Material Thickness																
$\frac{1}{8}$	8.35	9.75	10.5	11.6	12.8	13.9	15.1	23.2	9.11	10.6	11.4	12.7	13.9	15.2	16.5	25.3
$\frac{3}{16}$	12.5	14.6	15.7	17.4	19.2	20.9	22.6	34.8	13.7	15.9	17.1	19.0	20.9	22.8	24.7	38.0
$\frac{1}{4}$	16.7	19.5	20.9	23.2	25.5	27.8	30.2	46.4	18.2	21.3	22.8	25.3	27.9	30.4	32.9	50.6
$\frac{5}{16}$	20.9	24.4	26.1	29.0	31.9	34.8	37.7	58.0	22.8	26.6	28.5	31.6	34.8	38.0	41.2	63.3
$\frac{3}{8}$	25.1	29.2	31.3	34.8	38.3	41.8	45.3	69.6	27.3	31.9	34.2	38.0	41.8	45.6	49.4	75.9
$\frac{7}{16}$	29.2	34.1	36.6	40.6	44.7	48.7	52.8	81.2	31.9	37.2	39.9	44.3	48.8	53.2	57.6	88.6
$\frac{1}{2}$	33.4	39.0	41.8	46.4	51.1	55.7	60.4	92.8	36.5	42.5	45.6	50.6	55.7	60.8	65.9	101
$\frac{9}{16}$	37.6	43.9	47.0	52.2	57.5	62.6	67.9	104	41.0	47.8	51.3	57.0	62.7	68.3	74.1	114
$\frac{5}{8}$	41.8	48.7	52.3	58.0	63.9	69.6	75.5	116	45.6	53.2	57.0	63.3	69.7	75.9	82.3	127
$\frac{11}{16}$	45.9	53.6	57.5	63.8	70.2	76.6	83.0	128	50.1	58.5	62.7	69.6	76.6	83.5	90.5	139
$\frac{3}{4}$	50.1	58.5	62.7	69.6	76.6	83.5	90.5	139	54.7	63.8	68.4	75.9	83.6	91.1	98.8	152
$\frac{13}{16}$	54.3	63.3	67.9	75.4	83.0	90.5	98.1	151	59.2	69.1	74.1	82.3	90.6	98.7	107	165
$\frac{7}{8}$	58.5	68.2	73.1	81.2	89.4	97.5		162	63.8	74.4	79.8	88.6	97.5	106	115	177
$\frac{15}{16}$	62.7	73.1	78.4	87.0	95.8			174	68.3	79.7	85.5	94.9	104	114		190
1	66.8	78.0	83.6	92.8	102	111	121	186	72.9	85.1	91.2	101	111	122	132	203
$1\frac{1}{16}$	71.0	82.8	88.8	98.6				197	77.5	90.4	96.9	108	118			215
$1\frac{1}{8}$	75.2	87.7	94.1					209	82.0	95.7	103	114				228
$1\frac{3}{16}$	79.4	92.6	99.3					220	86.6	101	108					240
$1\frac{1}{4}$	83.5	97.5						232	91.1	106	114					253
$1\frac{5}{16}$	87.7							244	95.7	112						266
$1\frac{3}{8}$	91.9							255	100	117						278
$1\frac{7}{16}$	96.1							267	105							291
$1\frac{1}{2}$									109							304
$1\frac{9}{16}$									114							316

Note: This table not applicable to fasteners in friction-type connections.
F_y is the yield stress of the connected material; see AISC specification, Section 1.5.2.2.
F_p, the unit bearing stress, applies equally to conditions of single shear and enclosed bearing.
Values for thicknesses not listed may be obtained by multiplying the unlisted thickness by the value given for a 1-in thickness in the appropriate F_y column.
Values for F_y not listed may be obtained by multiplying the value given for $F_y = 100$ ksi by the unlisted F_y and dividing by 100.

Smith-Emery Test*

* This appendix is adapted from the Smith-Emery
Company, Los Angeles, California, 1975.

SUBJECT: Load tests on seismic snubber model Z-1011-1250.

SOURCE: Submitted to laboratory.

PROCEDURE: The seismic snubber was mounted in a test fixture in various positions and placed in a Baldwin Universal testing machine. A dial indicator was attached to the seismic snubber and movement recorded. See Figure C.1 for test position, X, Y, and Z.

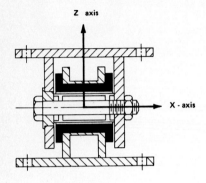

THIS UNIT MAY BE
SECURED TO EQUIPMENT
AND BUILDING STRUCTURE
BY BOLTING OR WELDING

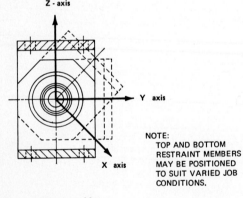

NOTE:
TOP AND BOTTOM
RESTRAINT MEMBERS
MAY BE POSITIONED
TO SUIT VARIED JOB
CONDITIONS.

FIG. C.1.   Seismic snubber.

REPORT OF TESTS

TEST NO. 1—X POSITION

Applied Load, lb	Movement, in	Observation
0	0.000	Neoprene touching bracket.
400	0.121	
800	0.228	
1250	0.342	
1600	0.430	
2000	0.522	
2500	0.628	Neoprene bushing bottomed out, and side plate on outer bracket began to deform.
3000	0.694	
3500	0.777	
4000	0.857	
4500	0.948	
5000	1.026	
5500	1.114	
6000	1.195	
6500	1.271	Inner bracket base plate began to deform around anchor bolt.
7000	1.358	
7500	1.464	
7900		Maximum load. One bottom anchor bolt sheared on inner bracket base plate.

Observation of seismic snubber after test:
1. Inner bracket base plate not deformed.
2. Center bolt deformed.
3. No cracking of welding on base plates.
4. Top neoprene bushing has small cracks present.

Test No. 2 — Y Position

Applied Load, lb	Movement, in	Observation
0	0.000	Neoprene touching bracket.
400	0.089	
800	0.162	
1000	0.196	Neoprene bushing bottomed out on bracket.
1250	0.230	
1600	0.284	
2000	0.363	
2500	0.457	
3000	0.518	
3750	0.603	
4000	0.637	
4500	0.689	Anchor bolts on outer bracket began to deform as clearance indicated under base plate.
5000	0.738	
5500	0.795	
6000	0.841	Inner bracket base plate began to deform.
6500	0.904	
7000	1.957	
7500	1.016	
8000	1.075	
8500	1.145	Outer bracket base plate began to deform.
9000	1.331	
9700		Maximum load. One outer bracket anchor bolt failed.

Observation of seismic snubber after test:
1. One anchor bolt failed.

Test No. 3 — Z Position

Applied Load, lb	Movement, in	Observation
0	0.000	
400	0.061	
800	0.134	
1200	0.195	
1600	0.247	
2000	0.293	Neoprene bushing bottomed out.
2400	0.312	
2800	0.344	
3200	0.380	
3600	0.396	
4000	0.421	
5000	0.487	Center bolt began to bend.
6000	0.541	
7000	0.581	Inner bracket base plate began to deform.
8000	0.628	Slight bow in inner bracket base plates at anchor bolt.
9000	0.668	
10000	0.709	
10700		Maximum load. Center bolt failed.

Observation of seismic snubber after test:
1. Inner bracket base plate slightly deformed at anchor bolts.
2. No deformation of welding.
3. Small crack in neoprene bushing.

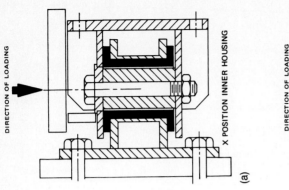

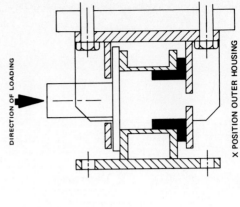

FIG. C.3. (a) X position of inner housing. (b) X position of outer housing.

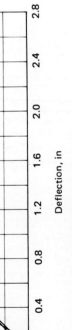

FIG. C.2. Test results showing movement of seismic snubber.

Uniform Building Code — Seismic Excerpts*

Earthquake Regulations

Sec. 2312. (a) **General.** Every building or structure and every portion thereof shall be designed and constructed to resist stresses produced by lateral forces as provided in this section. Stresses shall be calculated as the effect of a force applied horizontally at each floor or roof level above the base. The force shall be assumed to come from any horizontal direction.

Structural concepts other than set forth in this section may be approved by the building official when evidence is submitted showing that equivalent ductility and energy absorption are provided.

Where prescribed wind loads produce higher stresses, such loads shall be used in lieu of the loads resulting from earthquake forces.

(b) **Definitions.** The following definitions apply only to the provisions of this section:

BASE is the level at which the earthquake motions are considered to be imparted to the structure or the level at which the structure as a dynamic vibrator is supported.

BOX SYSTEM is a structural system without a complete vertical load-carrying space frame. In this system the required lateral forces are resisted by shear walls or braced frames as hereinafter defined.

BRACED FRAME is a truss system or its equivalent which is provided to resist lateral forces in the frame system and in which the members are subjected primarily to axial stresses.

DUCTILE MOMENT-RESISTING SPACE FRAME is a moment-resisting space frame complying with the requirements for a ductile moment-resisting space frame as given in Section 2312 (j).

ESSENTIAL FACILITIES—See Section 2312 (k).

LATERAL FORCE-RESISTING SYSTEM is that part of the structural system assigned to resist the lateral forces prescribed in Section 2312 (d) 1.

MOMENT-RESISTING SPACE FRAME is a vertical load-carrying space frame in which the members and joints are capable of resisting forces primarily by flexure.

SHEAR WALL is a wall designed to resist lateral forces parallel to the wall.

SPACE FRAME is a three-dimensional structural system without bearing walls, composed of interconnected members laterally supported so as to function as a complete self-contained unit with or without the aid of horizontal diaphragms or floor-bracing systems.

VERTICAL LOAD-CARRYING SPACE FRAME is a space frame designed to carry all vertical loads.

---

* This appendix is from the Uniform Building Code, 1979 edition.

(c) **Symbols and Notations.** The following symbols and notations apply only to the provisions of this section:

C = numerical coefficient as specified in Section 2312 (d) 1

C_p = numerical coefficient as specified in Section 2312 (g) and as set forth in Table 23-J

D = dimension of structure, ft, in direction parallel to applied forces

δ_i = deflection at level i relative to base, due to applied lateral forces, Σf_i, for use in formula (12-3)

F_i, F_n, F_x = lateral force applied to level i, n, or x, respectively

F_p = lateral forces on part of structure and in direction under consideration

F_t = that portion of V considered concentrated at top of structure in addition to F_n

f_i = distributed portion of total lateral force at level i for use in formula (12-3)

g = acceleration due to gravity

h_i, h_n, h_x = height, ft, above the base to level i, n, or x, respectively

I = occupancy importance factor as set forth in Table 23-K

K = numerical coefficient as set forth in Table 23-I [not included in this appendix]

Level i

l = level of structure referred to by subscript i

i = 1 designates first level above base

Level n

= that level which is uppermost in main portion of structure

Level x

= that level which is under design consideration

x = 1 designates first level above base

N = total number of stories above base to level n

S = numerical coefficient for site-structure resonance

T = Fundamental elastic period of vibration of building or structure, s, in direction under consideration

T_s = characteristic site period

V = total lateral force or shear at base

W = total dead load as defined in Section 2302 including partition loading specified in Section 2304 (d) where applicable

EXCEPTION: W shall be equal to the total dead load plus 25 percent of the floor live load in storage and warehouse occupancies. Where the design snow load is 30 lb/ft² or less, no part need be included in the value of W. Where the snow load is greater than 30 lb/ft², the snow load shall be included; however, where the snow load duration warrants, the building official may allow the snow load to be reduced up to 75 percent.

$w_i w_x$ = That portion of W which is located at or is assigned to level i or x, respectively

W_p = weight of portion of structure or nonstructural component

Z = numerical coefficient dependent upon zone as determined by Figures 1 to 3 in this chapter. For locations in zone 1, $Z = \frac{3}{16}$. For locations in zone 2, $Z = \frac{3}{8}$. For locations in zone 3, $Z = \frac{3}{4}$. For locations in zone 4, $Z = 1$.

(d) **Minimum Earthquake Forces for Structures.** Except as provided in Section 2312 (g) and (l), every structure shall be designed and constructed to resist minimum total lateral seismic forces assumed to act nonconcurrently in the direction of each of the main axes of the structure in accordance with the following formula:

$$V = ZIKCSW \qquad (12\text{-}1)$$

The value of K shall be not less than that set forth in Table 23-I. The value of C and S are as indicated hereafter except that the product of CS need not exceed 0.14.

The value of C shall be determined in accordance with the following formula:

$$C = \frac{1}{15\sqrt{T}} \qquad (12\text{-}2)$$

The value of C need not exceed 0.12.

The period T shall be established using the structural properties and deformational characteristics of the resisting elements in a properly substantiated analysis such as the following formula:

$$T = 2\pi \sqrt{\left(\sum_{i=1}^{n} \omega_i \delta_i^2 \right) \Big/ \left(g \sum_{i=1}^{n} f_i \delta_i \right)} \qquad (12\text{-}3)$$

where the values of f_i represent any lateral force

distributed approximately in accordance with the principles of formulas (12-5) to (12-7) or any other rational distribution. The elastic deflections δ_i shall be calculated using the applied lateral forces f_i.

In the absence of a determination as indicated above, the value of T for buildings may be determined by the following formula:

$$T = \frac{0.05h_n}{\sqrt{D}} \qquad (12\text{-}3A)$$

Or in buildings in which the lateral force-resisting system consists of ductile moment-resisting space frames capable of resisting 100 percent of the required lateral forces and such system is not enclosed by or adjoined by more rigid elements tending to prevent the frame from resisting lateral forces:

$$T = 0.10N \qquad (12\text{-}3B)$$

The value of S shall be determined by the following formulas, but shall be not less than 1.0:

For $T/T_s = 1.0$ or less

$$S = 1.0 + \frac{T}{T_s} - 0.5 \left(\frac{T}{T_N}\right)^2 \qquad (12\text{-}4)$$

For T/T_s greater than 1.0

$$S = 1.2 + 0.6 \frac{T}{T_s} - 0.3 \left(\frac{T}{T_s}\right)^2 \qquad (12\text{-}4A)$$

T in formulas (12-4) and (12-4A) shall be established by a properly substantiated analysis, but T shall be not less than 0.3 s.

The range of values of T_s may be established from properly substantiated geotechnical data, in accordance with UBC standard 23-1 except that T_s shall not be taken as less than 0.5 s nor more than 2.5 s. T_s shall be that value within the range of site periods, as determined above, that is nearest to T.

When T_s is not properly established, the value of S shall be 1.5.

EXCEPTION: WHERE T has been established by a properly substantiated analysis and exceeds 2.5 s, the value of S may be determined by assuming a value of 2.5 s for T_s.

(e) **Distribution of Lateral Forces. 1. Structures having regular shapes or framing systems.** The total lateral force V shall be distributed over the height of the structure in accordance with formulas (12-5) to (12-7).

$$V = F_t + \sum_{i=1}^{n} F_i \qquad (12\text{-}5)$$

The concentrated force at the top shall be determined according to the following formula:

$$F_t = 0.07TV \qquad (12\text{-}6)$$

F_t need not exceed 0.25 V and may be considered as 0 where T is 0.7 s or less. The remaining portion of the total base shear V shall be distributed over the height of the structure including level n according to the following formula:

$$F_x = \frac{(V - F_t)w_x h_x}{\sum_{i=1}^{n} w_i h_i} \qquad (12\text{-}7)$$

At each level designated as x, the force F_x shall be applied over the area of the building in accordance with the mass distribution on that level.

2. **Setbacks.** Buildings having setbacks wherein the plan dimension of the tower in each direction is at least 75 percent of the corresponding plan dimension of the lower part may be considered as uniform buildings without setbacks, provided other irregularities as defined in this section do not exist.

3. **Structures having irregular shapes or framing systems.** The distribution of the lateral forces in structures which have highly irregular shapes, large differences in lateral resistance or stiffness between adjacent stories, or other unusual structural features shall be determined considering the dynamic characteristics of the structure.

4. **Distribution of horizontal shear.** Total shear in any horizontal plane shall be distributed to the various elements of the lateral force-resisting system in proportion to their rigidities considering the rigidity of the horizontal bracing system or diaphragm.

Rigid elements that are assumed not to be part of the lateral force-resisting system may be incorporated into buildings provided that their effect on the action of the system is considered and provided for in the design.

5. **Horizontal torsional moments.** Provisions shall be made for the increase in shear resulting from the horizontal torsion due to an eccentricity between the center of mass and the center of

rigidity. Negative torsional shears shall be neglected. Where the vertical resisting elements depend on diaphragm action for shear distribution at any level, the shear-resisting elements shall be capable of resisting a torsional moment assumed to be equivalent to the story shear acting with an eccentricity of not less than 5 percent of the maximum building dimension at that level.

(f) **Overturning.** Every building or structure shall be designed to resist the overturning effects caused by the wind forces and related requirements specified in Section 2311 or the earthquake forces specified in this section, whichever governs.

At any level the incremental changes of the design overturning moment, in the story under consideration, shall be distributed to the various resisting elements in the same proportion as the distribution of the shears in the resisting system. Where other vertical members are provided which are capable of partially resisting the overturning moments, a redistribution may be made to these members if framing members of sufficient strength and stiffness to transmit the required loads are provided.

Where a vertical resisting element is discontinuous, the overturning moment carried by the lowest story of that element shall be carried down as loads to the foundation.

(g) **Lateral Force on Elements of Structures and Nonstructural Components.** Parts or portions of structures, nonstructural components, and their anchorage to the main structural system shall be designed for lateral forces in accordance with the following formula:

$$F_p = ZIC_pW_p \tag{12-8}$$

The values of C_p are set forth in Table 23-J. The value of the I coefficient shall be the value used for the building.

EXCEPTIONS: 1. The value of I for panel connectors shall be as given in Section 2312 (j) 3 C.
2. The value of I for anchorage of machinery and equipment required for life safety systems shall be 1.5.

The distribution of these forces shall be according to the gravity loads pertaining thereto.

For applicable forces on diaphragms and connections for exterior panels, refer to Sections 2312 (j) 2 D and 2312 (j) 3 C.

(h) **Drift and Building Separations.** Lateral deflections or drift of a story relative to its adjacent stories shall not exceed 0.005 times the story height unless it can be demonstrated that greater drift can be tolerated. The displacement calculated from the application of the required lateral forces shall be multiplied by $1.0/K$ to obtain the drift. The ratio $1.0/K$ shall be not less than 1.0.

All portions of structures shall be designed and constructed to act as an integral unit in resisting horizontal forces unless separated structurally by a distance sufficient to avoid contact under deflection from seismic action or wind forces.

(i) **Alternate Determination and Distribution of Seismic Forces.** Nothing in Section 2312 shall be deemed to prohibit the submission of properly substantiated technical data for establishing the lateral forces and distribution by dynamic analyses. In such analyses the dynamic characteristics of the structure must be considered.

(j) **Structural Systems. 1. Ductility requirements.** A. All buildings designed with a horizontal force factor $K = 0.67$ or 0.80 shall have ductile moment-resisting space frames.

B. Buildings more than 160 ft in height shall have ductile moment-resisting space frames capable of resisting not less than 25 percent of the required seismic forces for the structure as a whole.

EXCEPTION: Buildings more than 160 ft in height in seismic zones 1 and 2 may have concrete shear walls designed in accordance with Section 2627 or braced frames designed in conformance with Section 2312 (j) 1 G of this code in lieu of a ductile moment-resisting space frame, provided a K value of 1.00 or 1.33 is utilized in the design.

C. In seismic zones 2 to 4, all concrete space frames required by design to be part of the lateral force-resisting system and all concrete frames located in the perimeter line of vertical support shall be ductile moment-resisting space frames.

EXCEPTION: Frames in the perimeter line of the vertical support of buildings designed with shear walls taking 100 percent of the design lateral forces need only conform with Section 2312 (j) 1 D.

D. In seismic zones 2 to 4, all framing elements not required by design to be part of the lateral force-resisting system shall be investigated and shown to be adequate for vertical load-carrying capacity and induced moment due to $3/K$ times the distortions resulting from the code-required lateral forces. The rigidity of other elements shall be considered in accordance with Section 2312 (e) 4.

E. Moment-resisting space frames and ductile moment-resisting space frames may be enclosed by or adjoined by more rigid elements that would tend to prevent the space frame from resisting lateral forces where it can be shown that the action or failure of the more rigid elements will not impair the vertical and lateral load resisting ability of the space frame.

F. Necessary ductility for a ductile moment-resisting space frame shall be provided by a frame of structural steel with moment-resisting connections (complying with Section 2722 for buildings in seismic zones 3 and 4 or Section 2723 for buildings in seismic zones 1 and 2) or by a reinforced concrete frame (complying with Section 2626 for buildings in seismic zones 3 and 4 or Section 2625 for buildings in seismic zones 1 and 2).

> **EXCEPTION:** Buildings with ductile moment-resisting space frames in seismic zones 1 and 2 having an importance factor I greater than 1.0 comply with Section 2626 or 2722.

G. In seismic zones 3 and 4 and for buildings having an importance factor I greater than 1.0 located in seismic zone 2, all members in braced frames shall be designed for 1.25 times the force determined in accordance with Section 2312 (d). Connections shall be designed to develop the full capacity of the members or shall be based on the above forces without the one-third increase usually permitted for stresses resulting from earthquake forces.

Braced frames in buildings shall be composed of axially loaded bracing members of A36, A440, A441, A501, A572 (except grades 60 and 65) or A588 structural steel or reinforced concrete members conforming to the requirements of Section 2627.

H. Reinforced concrete shear walls for all buildings shall conform to the requirements of Section 2627.

I. In structures where $K = 0.67$ and $K = 0.80$, the special ductility requirements for structural steel or reinforced concrete specified in Section 2312 (j) 1 F shall apply to all structural elements below the base that are required to transmit to the foundation the forces resulting from lateral loads.

2. **Design requirements.** A. **Minor alterations.** Minor structural alterations may be made in existing buildings and other structures, but the resistance to lateral forces shall be not less than that before such alterations were made, unless the

building as altered meets the requirements of this section.

B. **Reinforced masonry or concrete.** All elements within structures located in seismic zones 2 to 4 that are of masonry or concrete shall be reinforced so as to qualify as reinforced masonry or concrete under the provisions of Chapters 24 and 26. Principal reinforcement in masonry shall be spaced 2 ft maximum on center in buildings using a moment-resisting space frame.

C. **Combined vertical and horizontal forces.** In computing the effect of seismic force in combination with vertical loads, gravity load stresses induced in members by dead load plus design live load, except roof live load, shall be considered. Consideration should also be given to minimum gravity loads acting in combination with lateral forces.

D. **Diaphragms.** Floor and roof diaphragms and collectors shall be designed to resist the forces determined in accordance with the following formula:

$$F_{px} = \frac{\sum\limits_{i=1}^{n} F_l}{\sum\limits_{i=1}^{n} w_l} w_{px} \qquad (12\text{-}9)$$

Where F_l = lateral force applied to level l
  w_l = portion of W at level l
  w_{px} = weight of diaphragm and elements tributary thereto at level x, including 25 percent of floor live load in storage and warehouse occupancies

The force F_{px} determined from formula (12-9) need not exceed $0.30ZIw_{px}$.

When the diaphragm is required to transfer lateral forces from the vertical resisting elements above the diaphragm to other vertical resisting elements below the diaphragm due to offsets in the placement of the elements or to changes in stiffness in the vertical elements, these forces shall be added to those determined from formula (12-9).

However, in no case shall lateral force on the diaphragm be less than $0.14ZIw_{px}$.

Diaphragms supporting concrete or masonry walls shall have continuous ties between diaphragm chords to distribute, into the diaphragm, the anchorage forces specified in this chapter. Added chords may be used to form subdiaphragms to transmit the anchorage forces to the main cross ties. Diaphragm deformations shall be considered in the design of the supported walls. See Section 2312 (j) 3 A for special anchorage requirements of wood diaphragms.

3. **Special requirements. A. Wood diaphragms providing lateral support for concrete or masonry walls.** Where wood diaphragms are used to laterally support concrete or masonry walls the anchorage shall conform to Section 2310. In zones 2 to 4, anchorage shall not be accomplished by use of toenails or nails subjected to withdrawal, nor shall wood framing be used in cross-grain bending or cross-grain tension.

B. **Pile caps and caissons.** Individual pile caps and caissons of every building or structure shall be interconnected by ties, each of which can carry by tension and compression a minimum horizontal force equal to 10 percent of the larger pile cap or caisson loading, unless it can be demonstrated that equivalent restraint can be provided by other approved methods.

C. **Exterior elements.** Precast or prefabricated nonbearing, nonshear wall panels or similar elements that are attached to or enclose the exterior shall be designed to resist the forces determined from formula (12-8) and shall accommodate movements of the structure resulting from lateral forces or temperature changes. The concrete panels or other similar elements shall be supported by means of cast-in-place concrete or mechanical connections and fasteners in accordance with the following provisions:

Connections and panel joints shall allow for a relative movement between stories of not less than 2 times story drift caused by wind or $(3.0/K)$ times the calculated elastic story displacement caused by required seismic forces, or ½ in, whichever is greater. Connections to permit movement in the plane of the panel for story drift shall be properly designed sliding connections using slotted or oversized holes or may be connections that permit movement by bending of steel or other connections providing equivalent sliding and ductility capacity.

Bodies of connectors shall have sufficient ductility and rotation capacity so as to preclude fracture of the concrete or brittle failures at or near welds.

The body of the connector shall be designed for one and one-third times the force determined by formula (12-8). Fasteners attaching the connector to the panel or the structure such as bolts, inserts, welds, dowels, etc., shall be designed to ensure ductile behavior of the connector or shall be designed for 4 times the load determined from formula (12-8).

Fasteners embedded in concrete shall be attached to or hooked around reinforcing steel or otherwise terminated so as to effectively transfer forces to the reinforcing steel.

The value of the coefficient I shall be 1.0 for the entire connector assembly in formula (12/8).

(k) **Essential Facilities.** Essential facilities are those structures or buildings that must be safe and usable for emergency purposes after an earthquake in order to preserve the health and safety of the general public. Such facilities shall include but not be limited to

1. Hospitals and other medical facilities having surgery or emergency treatment areas

2. Fire and police stations

3. Municipal government disaster operation and communication centers deemed to be vital in emergencies.

The design and detailing of equipment that must remain in place and be functional following a major earthquake shall be based upon the requirements of Section 2312 (g) and Table 23-J. In addition, their design and detailing shall consider effects induced by structure drifts of not less than $2.0/K$ times the story drift caused by required seismic forces nor less than the story drift caused by wind. Special consideration shall also be given to relative movements at separation joints.

(l) **Earthquake-recording Instrumentations.** For earthquake recording instrumentations, see Appendix, Section 2312 (l).

Table 23-A Uniform and Concentrated Loads

Use or Occupancy Category	Description	Uniform Load[a]	Concentrated Load	Use or Occupancy Category	Description	Uniform Load[a]	Concentrated Load
1. Armories		150	0	9. Offices		50	2000[f]
2. Assembly areas[b] and auditoriums and balconies therewith	Fixed seating areas	50	0	10. Printing plants	Press rooms	150	2500[f]
	Movable seating and other areas	100	0		Composing and linotype rooms	100	2000[f]
	Stage areas and enclosed platforms	125	0	11. Residential[g]		40	0[d]
3. Cornices, marquees, and residential balconies		60	0	12. Rest rooms[h]			
				13. Reviewing stands, grand stands, and bleachers		100	0
4. Exit facilities[c]		100	0[d]	14. Roof deck	Same as area served or for the type of occupancy accommodated		
5. Garages	General storage and/or repair	100	e				
	Private pleasure car storage	50	e	15. Schools	Classrooms	40	1000[f]
6. Hospitals	Wards and rooms	40	1000[f]	16. Sidewalks and driveways	Public access	250	e
7. Libraries	Reading rooms	60	1000[f]	17. Storage	Light	125	
	Stack rooms	125	1500[f]		Heavy	250	
8. Manufacturing	Light	75	2000[f]	18. Stores	Retail	75	2000[f]
	Heavy	125	3000[f]		Wholesale	100	3000[f]

[a] See Section 2306 for live load reductions.

[b] Assembly areas include such occupancies as dance halls, drill rooms, gymnasiums, playgrounds, plazas, terraces, and similar occupancies which are generally accessible to the public.

[c] Exit facilities shall include such uses as corridors serving an occupant load of 10 or more persons, exterior exit balconies, stairways, fire escapes, and similar uses.

[d] Individual stair treads shall be designed to support a 300-lb concentrated load placed in a position which would cause maximum stress. Stair stringers may be designed for the uniform load set forth in the table.

[e] See Section 2304 (c), second paragraph, for concentrated loads.

[f] See Section 2304 (c), first paragraph, for area of load application.

[g] Residential occupancies include private dwellings, apartments, and hotel guest rooms.

[h] Rest room loads shall be not less than the load for the occupancy with which they are associated but need not exceed 50 lb/ft².

Table 23-B Special Loads[a]

Use		Vertical Load	Lateral Load	Use		Vertical Load	Lateral Load
Category	Description	(lb/ft² Unless Otherwise Noted)		Category	Description	(lb/ft² Unless Otherwise Noted)	
1. Construction, public access at site (live load)	Walkway see Section 4406	150		4. Ceiling framing (live load)(Cont.)	All uses except over stages	10[e]	
	Canopy see Section 4407	150		5. Partitions and interior walls, see Section 2309 (live load)			5
2. Grandstands, reviewing stands, and bleachers (live load)	Seats and footboards	120[b]	See footnote C	6. Elevators and dumbwaiters (dead and live load)		2 × total loads[f]	
3. Stage accessories, see Section 3902 (live load)	Gridirons and fly galleries	75		7. Cranes (dead and live load[g])	Total load including impact increase	1.25 × total load[g]	0.10 × total load[h]
	Loft block wells[d]	250	250	8. Balcony railings, guard rails, and handrails	Exit facilities serving an occupant load greater than 50		50[i]
	Head block wells and sheave beams[d]	250	250		Other		20[i]
4. Ceiling framing (live load)	Over stages	20		9. Storage racks	Over 8 ft high	Total loads[i]	See Table 23-J

[a] The tabulated loads are minimum loads. Where other vertical loads required by this code or required by the design would cause greater stresses they shall be used.

[b] Pounds per lineal foot.

[c] Lateral sway bracing loads of 24 lb/ft parallel and 10 lb/ft perpendicular to seat and footboards.

[d] All loads are in pounds per lineal foot. Head block wells and sheave beams shall be designed for all loft block well loads tributary thereto. Sheave blocks shall be designed with a factor of safety of five.

[e] Does not apply to ceilings which have sufficient total access from below, such that access is not required within the space above the ceiling. Does not apply to ceilings if the attic areas above the ceiling are not provided with access. This live load need not be considered acting simultaneously with other live loads imposed upon the ceiling framing or its supporting structure.

[f] Where Appendix Chapter 51 has been adopted, see reference standard cited therein for additional design requirements.

[g] The impact factors included are for cranes with steel wheels riding on steel rails. They may be modified if substantiating technical data acceptable to the building official is submitted. Live loads on crane support girders and their connections shall be taken as the maximum crane wheel loads. For pendant-operated traveling crane support girders and their connections, the impact factors shall be 1.10.

[h] This applies in the direction parallel to the runway rails (longitudinal). The factor for forces perpendicular to the rail is 0.20 × the transverse traveling loads (trolley, cab, hooks, and lifted loads). Forces shall be applied at top of rail and may be distributed among rails of multiple rail cranes and shall be distributed with due regard for lateral stiffness of the structures supporting these rails.

[i] A load per lineal foot to be applied horizontally at right angles to the top rail.

[j] Vertical members of storage racks shall be protected from impact forces of operating equipment or racks shall be designed so that failure of one vertical member will not cause collapse of more than the bay or bays directly supported by that member.

Table 23-J **Horizontal Force Factor C_p for Elements of Structures and Nonstructural Components**

Part or Portion of Buildings	Direction of Horizontal Force	Value of C_p[a]	Part or Portion of Buildings	Direction of Horizontal Force	Value of C_p[a]
1. Exterior bearing and nonbearing walls, interior bearing walls and partitions, interior nonbearing walls and partitions—see also Section 2312 (j) 3 C. Masonry or concrete fences over 6 ft high	Normal to flat surface	0.3[b]	4. When connected to, part of, or housed within a building: a. Penthouses, anchorage, and supports for chimneys and stacks and tanks, including contents b. Storage racks with upper storage level at more than 8 ft in height, plus contents[c]		
2. Cantilever elements: a. Parapets	Normal to flat surfaces	0.8			
b. Chimneys or stacks	Any direction		5. Suspended ceiling framing systems (applies to seismic zones 2 to 4 only)	Any direction	0.3[d]
3. Exterior and interior ornamentations and appendages	Any direction	0.8	6. Connections for prefabricated structural elements other than walls, with force applied at center of gravity of assembly	Any direction	0.3[e]

[a] C_p for elements laterally self-supported only at the ground level may be two-thirds of value shown.

[b] See also Section 2309 (b) for minimum load and deflection criteria for interior partitions.

[c] W_p for storage racks shall be the weight of the racks plus contents. The value of C_p for racks over two storage support levels in height shall be 0.24 for the levels below the top two levels. In lieu of the tabulated values steel storage racks may be designed in accordance with UBC standard 27-11.

Where a number of storage rack units are interconnected so that there are a minimum of four vertical elements in each direction on each column line designed to resist horizontal forces, the design coefficients may be as for a building with K values from Table 23-I, $CS = 0.2$ for use in the formula $V = ZIKCSW$ and W equal to the total dead load plus 50 percent of the rack-rated capacity. Where the design and rack configurations are in accordance with this paragraph, the design provisions in UBC standard 27-11 do not apply. For essential facilities and life safety systems, the design and detailing of equipment that must remain in place and be functional following a major earthquake shall consider drifts in accordance with Section 2312 (k).

[d] Ceiling weight shall include all light fixtures and other equipment that is laterally supported by the ceiling. For purposes of determining the lateral force, a ceiling weight of not less than 4 lb/ft² shall be used.

[e] The force shall be resisted by positive anchorage and not by friction.

Table 23-K **Values for Occupancy Importance Factor I**

Type of Occupancy	
Essential Facilities*	1.5
Any building where the primary occupancy is for assembly use for more than 300 persons (in one room)	1.25
All others	1.0

The first row value:

Type of Occupancy	1

* See Section 2312 (k) for definition and additional requirements for essential facilities.

Register 74— Safety of Construction of Hospitals*

45081. Applications, Procedures, and Approvals. (a) **Applications (General).** Before letting any contract for the construction of, reconstruction of, alteration of, or addition to any hospital building, the governing board or authority of the hospital shall submit an application to the department for approval of plans and specifications and shall have had and obtained the written approval thereof by the department, describing the scope of work included and any special conditions under which approval is given. Refer to Section 45260 regarding incremental bidding.

(b) **Submission of documents shall be made in three stages:** namely, site data, preliminary plans and outline specifications, and working drawings and final specifications. The first submission shall be submitted in three copies, accompanied by three copies of an "Application for Approval of Plans and Specifications", a filing fee deposit, and one copy of "Application for Health Facility License".

(1) **Site Data.** Geologic and earthquake engineering report. This report will be submitted by the Department to the State Division of Mines and Geology for review and shall provide the following data:

(A) A geologic and earthquake engineering report(s) containing data on an assessment of the nature of the site and potential for earth-

quake damage based upon geologic, foundation, and earthquake engineering investigation(s) made jointly but with appropriately divided responsibility by a California certified engineering geologist (engineering geology report) and a registered civil engineer (earthquake engineering report) shall be submitted in triplicate. The geologic report shall not be considered to contain design criteria but only scientific data to be used in the earthquake engineering evaluation. The report shall include but shall not be limited to the following:

1. Geologic investigation.

2. Evaluation of known active and potentially active faults, both regional and local.

3. Evaluation of slope stability and liquefaction potential.

4. Evaluation of any other potential geologic hazards, including those described in CDMG Notes #37 and Special Publication 42, 1973, except that estimates of ground motions (including acceleration data) need not be included in this report.

(B) The geologic and earthquake engineering report(s) will be required for all proposed construction, whether it is to be designed by methods A, B, or C, as set forth in Section T-17-2314(d), Title 24, California Administrative Code, with the exception of one-story buildings of type V construction and 4000 ft² or

---

* This appendix is from Title 17, Safety of Construction of Hospitals, State of California, 1974.

less in floor area, or for nonstructural, associated structural, or nonrequired structural alterations; structural repairs for other than earthquake damage; or incidental structural additions or alterations.

(C) A previous report for a specific site may be resubmitted provided that a reevaluation is made by the geotechnical team and the report is found to be appropriate.

(2) **Preliminary Drawings and Outline Specifications.** These documents shall be submitted in triplicate and all the documents listed in subsection (b) (1) (A) above shall have been submitted. These documents will be reviewed in the department as well as in the Office of the State Fire Marshal for guidance relative to Titles 17 and 19, California Administrative Code, respectively, and shall provide the following data:

(A) Preliminary drawings of the proposed building, any existing building on the site, and the relationships of various departments or services existing and proposed shall be shown. The name of each room (denoting its function) shall be noted on the floor plans. The drawings shall be at a legible scale and shall include:

1. Plot plan showing roads, courses and distances of property lines, existing buildings, proposed buildings, parking areas, sidewalks, topography, and any easements of record.

2. Plan of each floor, including the basement or ground floor.

3. Elevations of all sides and relevant sections.

4. Location of fixed and major movable equipment.

5. At the option of the engineer, the analysis for ground shaking for use with method A, as defined in Section (3) (A) below, may be submitted for review with the preliminary submittals.

(B) Outline specifications that provide a general description of the construction, including finishes and the type of mechanical and electrical systems.

(C) Design program briefly explaining functions and services of the proposed facility and including:

1. Classification or type of facility

2. Number and type of patients to be accommodated

3. Types of disabilities and conditions to be served

(3) **Working Drawings, Final Specifications, and Reports.** These documents shall be submitted in triplicate together with balance of filing fee and will be reviewed by the Department of General Services for structural safety, by the State Fire Marshal for fire safety, and by the department for general architectural, mechanical, and electrical details, and shall include the following information:

(A) **Geotechnic Report.** A detailed soil and foundation engineering report containing design criteria related to the nature and extent of foundation materials, groundwater conditions, slope stability, etc., shall be submitted for review. The report shall contain the results of analyses of problem areas identified in the geologic and earthquake engineering report(s). A report shall be prepared by the geotechnic engineer with the advice of the engineering geologist, if appropriate, in which case, both the engineer and the geologist shall sign the report(s).

Analysis of ground shaking for a specific site shall be required for method A, only. Where method A is used, the recommended earthquake ground motions for the site shall be reported in one or more of the following formats, depending on the needs of the structural engineer. Both the maximum credible earthquake and the function-basis earthquake, as defined in Section 45004 of these regulations, shall be reported:

1. Ground motion

2. Structural response motion

3. Elastic structural response spectra

4. Time-history plot of predicted ground motion at the site

5. Other data—standard seismicity

(B) Architectural drawings which shall include:

1. Plot plan.

2. Floor plans.

3. Roof plans.

4. Elevations and sections.

5. Necessary details.

6. Schedule of finishes, doors, and windows.

7. Location and identifying data on items of fixed equipment or major movable equipment such as autoclaves, sterilizers, kitchen equipment, laboratory equipment, x-ray equipment, and cubicle curtains. Anchorage details of all

fixed items shall be detailed as recommended by the structural engineer.

(C) Structural drawings which shall include:

1. Plans of foundations, floors, roofs, and any intermediate levels showing a complete design with sizes, sections, and relative location of the various members and a schedule of beams, girders, and columns. Assumed soil-bearing pressures and type of material shall be shown on foundation plans.

2. Details of all special connections, assemblies, and expansion joints.

Structural drawings shall be accompanied by computations, stress diagrams, and other pertinent data and shall be so complete that calculations for individual structural members can be readily interpreted.

(D) Mechanical drawings shall show the complete heating, ventilating, air-conditioning, and plumbing systems and details showing methods for fastening equipment to the structure to resist seismic forces and shall include:

1. Radiators and steam-heated equipment, such as sterilizers, autoclaves, warmers, and steam tables

2. Heating and steam mains, including branches with pipe sizes

3. Sizes, types, and heating surfaces of boilers and furnaces

4. Pumps, tanks, boiler breeching and piping, and boiler room accessories

5. Air-conditioning systems with refrigerators, water and refrigerant piping, and ducts

6. Exhaust and supply ventilating systems showing duct sizes with steam or water connections and piping

7. Size and elevation of street sewer, house sewer, house drains, street water main, and water service into the building

8. Location and size of soil, waste, and vent stacks with connections to house drains, fixtures, and equipment

9. Size and location of hot, cold, and circulation water mains, branches and risers from the service entrance, and tanks

10. Riser diagram or some other acceptable method to show all plumbing stacks with vents, water risers, and fixture connections for multistory buildings

11. Gas, oxygen, and special connections

12. Fire-extinguishing equipment—sprinklers, wet and dry standpipes, and fire extinguishers

13. Plumbing fixtures and fixtures that require water and drain connection

(E) Electrical drawings that show the complete electrical system and details showing methods for fastening equipment to the structure to resist seismic forces and shall include:

1. Electrical service entrance with service switches, service feeds to the public service feeders, and characteristics of the light and power currents

2. Transformers and their connections, if located in the building or on the site

3. Plan and diagram showing main switchboard, power panels, light panels, and equipment

4. Feeder and conduit sizes with schedule of feeder breakers or switches

5. Light outlets, receptacles, switches, power outlets and circuits, and isolated electrical system

6. Telephone layout

7. Nurses' call system with outlets for beds, duty stations, door signal lights, and annunciators

8. Fire alarm systems with stations, sounding devices, and control boards

9. Emergency lighting system when required with outlets, transfer switch, source of supply, feeders, and circuits

(F) Architectural, structural, mechanical, and electrical specifications shall fully describe (except where fully indicated and described on the drawings) the materials, workmanship and the kind, sizes, capacities, finishes, and other characteristics of all materials, products, articles, and devices.

(G) Addition to or alterations and repairs of existing structures shall include:

1. Types of activities within the existing buildings, including distribution

2. Type of construction of existing buildings and number of stories

3. Plans and details showing attachment of new construction to existing structural, mechanical, and electrical systems.

(H) A title block or strip on each sheet of the preliminary and of the working drawings shall state the following:

1. Name and address of the architect or engineer

2. Name and address of the project

3. Number or letter of each sheet

4. Date of preparation of each sheet and the date of revision, if any

5. The scale of each drawing or detail

(I) The north point of reference and the location dimensions or reference dimensions of the building with respect to the site boundaries shall be shown on all plot plans and on all floor plans where applicable.

(J) Unless working drawings and specifications are submitted to the department within 1 year of the report on preliminary drawings, such drawings and an "Application for Approval of Plans and Specifications" shall be resubmitted with necessary fee deposit for review.

(K) The written approval of plans and specifications shall be void as to any portion of the work not let under a construction contract within 1 year of the date of written approval—except when incremental bidding and construction has been approved by the department. Reinstatement of such approval will be granted subject to compliance of drawings with current code requirements. No new fee will be charged if plans have not been revised except as may be required by the department. No reinstatement of approval will be granted after 4 years from the date of the written approval.

(L) Routine maintenance and repairs normally done by the maintenance staff do not require prior approval by the department but shall be done in compliance with the applicable provisions of this code.

(M) Drawings and specifications shall comply with the provisions of these regulations and such applicable requirements of local authorities having jurisdiction. Alterations that do not affect the structural or fire safety features of the building shall be excluded from the requirement of structural review.

(N) **Signatures Required.** All plans and specifications shall bear the signature of the architect or engineer in responsible charge of design. When responsibility for a portion of the work has been delegated, the plans and specifications covering that portion of design shall also bear the signature of the architect or engineer delegated such responsibility.

(O) **Voidance of Application.** Any change, erasure, alteration, or modification of any drawing or specification bearing the identification stamp "Approved" of the department or the Department of General Services automatically voids the approval of the application. However, the written approval of plans may be extended to include revised and/or additional plans and specifications after submission for review and approval thereof.

(P) **Approval of Application.** The approval of the "Application for Approval of Plans and Specifications" constitutes the written approval of plans as to safety of design and construction required before any valid contract may be executed. One set of the plans and specifications bearing the stamps of approval shall be issued to the department and to the Department of General Services before the written approval of plans and specifications is issued.

History: 1. Amendment filed 9-20-73; effective thirtieth day thereafter (Register 73, No. 38).
2. Amendment of subsection (b)(1)(C) filed 4-30-74; effective thirtieth day thereafter (Register 74, No. 18).
3. Amendment filed 7-11-74; effective thirtieth day thereafter (Register 74, No. 28).
4. Amendment filed 12-24-74; effective thirtieth day thereafter (Register 74, No. 52).

State of California — Building Standards — Health Facilities

DIVISION T22.   OFFICE OF STATEWIDE HEALTH
PLANNING AND DEVELOPMENT

CHAPTER 2.   STRUCTURAL DESIGN, MATERIALS AND DETAILS
OF CONSTRUCTION

Article
1.   General Design Requirements (UBC Chapter 23)
2.   Masonry (UBC Chapter 24)
3.   Wood (UBC Chapter 25)
4.   Concrete (UBC Chapter 26)
5.   Steel and Iron (UBC Chapter 27)
6.   Aluminum (UBC Chapter 28)
7.   Excavations, Foundations, and Retaining Walls
          (UBC Chapter 29)
8.   Veneer (UBC Chapter 30)
9.   Roof Construction and Covering (UBC Chapter
          32)
10.   Masonry or Concrete Chimneys, Fireplaces,
          and Barbecues (UBC Chapter 37)
11.   Installation of Wall and Ceiling Coverings
          (UBC Chapter 47)
12.   Glass and Glazing (UBC Chapter 54)

DETAILED ANALYSIS

Article 1.   General Design Requirements (UBC Chapter 23)
Section
T22-94201   (UBC 2301). Scope
T22-94203   (UBC 2303). Design Methods
T22-94205   (UBC 2304). Floor Designs

T22-94207   (UBC 2305). Roof Design
T22-94209   (UBC 2307). Deflection
T22-94211   (UBC 2308). Special Design
T22-94213   (UBC 2309). Walls and Structural Framing
T22-94215   (UBC 2312). Earthquake Regulations

Article 2.   Masonry (UBC Chapter 24)

T22-94215   (UBC 2312). Earthquake Regulations.

(a) General.   Every hospital building or structure and every portion thereof including the nonstructural components shall be designed and constructed to resist stresses produced by lateral forces as provided in this section. Stresses shall be calculated as the effect of a force applied horizontally at each floor or roof level above the base. The force shall be assumed to come from any horizontal direction.

(1) Structural concepts other than set forth in this section may be approved by the department when evidence is submitted showing that equivalent ductility and energy absorption are provided.

(2) When the design of a structure, due to the unusual configuration of the structure or parts of the structure, does not provide at least the same safeguard against earthquakes as provided by the applicable portions of this section when applied in the design of a similar structure of customary configuration, framing and assembly of materials, the department shall withhold its approval.

Table T22-23J — Part A (UBC Table 23-J)   Horizontal Force Factor C_p for Elements of Structures[a]

Part of Portion of Buildings	Direction of Force	Value of C_p
1. Exterior bearing and nonbearing walls, interior bearing walls and partitions, interior nonbearing walls and partitions, and masonry or concrete fences	Normal to flat surface	0.20[b]
2. Cantilever parapet	Normal to flat surface	1.00
3. When part of a building, cantilever walls above the ground floor (except parapets)	Normal to flat surface	0.30
4. Floors and roofs acting as diaphragms	Any direction	0.12[c]
5. Penthouses (except where framed by an extension of the building space frame), towers, chimneys, and smokestacks	Any direction	0.20
6. Exterior and interior ornamentations and appendages	Any direction	1.00
7. Connections for exterior panels or for elements complying with Section T22-94215 (j) (3) (C)	Any direction	2.00
8. Connections for prefabricated structural elements other than walls, with force applied at center of gravity of assembly	Any direction	0.30[d]

[a] N. B. See Section T22-94215 (g) for use of C_p in design of elements of structures in accordance with the following formula:

$$F_p = ZIC_pSW_p \qquad (12\text{-}8)$$

[b] See also UBC Section 2309 (b) for minimum load and deflection criteria for interior partitions.

[c] Floors and roofs acting as diaphragms shall be designed for a minimum force resulting from a C_p of 0.12 applied to w, unless a greater force results from the distribution of lateral forces in accordance with UBC Section 2312 (e).

[d] The W_p shall include 25 percent of the floor live load in storage and warehouse occupancies.

Table T22-23-J — PART B (UBC Table 23-J)   Horizontal Force Factor C_p for Anchorage of Nonstructural Components[a]

Category	Direction of Force	Value of C_p[b,c] $I = 1.0$	Value of C_p[b,c] $I = 1.5$
1. Mechanical and electrical components a. All mechanical and electrical equipment or machinery[d]	Any direction	0.30[e]	0.50[e]
b. Tanks (plus contents) and support systems			
c. Essential communication equipment and emergency power equipment such as engine generators, battery racks, and fuel tanks necessary for operation of such equipment[f]		0.75[e]	0.75[e]
d. Piping, electrical conduit, cable trays, and air handling ducts[g]		0.50	0.50
2. Hospital equipment[g] when permanently attached to building utility services such as surgical, morgues and recovery room fixtures, radiology room equipment, medical gas containers, food service fixtures, essential laboratory equipment, TV supports	Any direction	0.30[e]	0.50[e]
3. a. Storage racks with the upper storage level more than 5 ft in height (plus contents)[h,i,j]	Any direction	0.30[e]	0.50[e]
b. Floor supported cabinets and bookstacks more than 5 ft in height (plus contents)[h,i]			

[a] N. B. See Section T22-94215 (m) for use of C_p for design of anchorage on nonstructural components in accordance with the following formula:

$$F_p = ZC_pW_p \tag{12-9}$$

Welded, bolted or other intermittent connections such as inserts for anchorage of nonstructural components shall not be allowed the one-third increase in allowable stress permitted in UBC Section 2303 (d).

[b] C_p may be two-thirds of value shown for components mounted on foundations at grade or on floor slabs on earth subgrade.

C_p shall be not less than the ratio of F_x/W_x for floor or roof level under consideration. Where a dynamic analysis is used in the design of the building, the forces so determined may be used in the design of the elements or components with appropriate resistance criteria.

[c] For flexible and flexibly mounted equipment and machinery (with fundamental period of vibration greater than 0.05 s), the appropriate values of C_p shall be determined with consideration given to both the dynamic properties of the equipment and machinery and to the buildings or structure in which it is placed, but shall not be less than the listed values. For closely restrained flexible mountings with resilient snubbers, a factor of two times the values shown shall be used except where more appropriate values can be determined by a dynamic analysis.

[d] Equipment or machinery shall include such items as tanks, boilers, chillers, pumps, motors, air-handling units, cooling towers, transformers, switch gear, and control panels, which is part of the building service systems and which is required for the continued function of the building.

[e] The component anchorage shall be designed for the horizontal force F_p, acting simultaneously with a vertical seismic force equal to one-third of the horizontal force F_p.

[f] Emergency equipment should be located where there is the least likelihood of damage due to earthquake. Such equipment should be located at ground level and where it can be easily maintained to assure its operation during an emergency.

[g] Seismic restraints may be omitted from the following installations:

a. Gas piping less than 1-in inside diameter.

b. Piping in boiler and mechanical equipment rooms less than 1¼-in inside diameter.

c. All other piping less than 2½-in inside diameter.

d. All piping suspended by individual hangers 12 in or less in length from the top of pipe to the bottom of the support for the hanger.

e. All electrical conduit less than 2½-in inside diameter.

f. All rectangular air-handling ducts less than 6 ft² in cross-sectional area.

g. All round air-handling ducts less than 28 in in diameter.

h. All ducts suspended by hangers 12 in or less in length from the top of the duct to the bottom of the support for the hanger.

[h] Floor-supported storage racks, cabinets, or book stacks not more than 8 ft in height need not be anchored if the width of the

Category	Direction of Force	Value of $C_p^{b,c}$	
		$I = 1.0$	$I = 1.5$
4. Wall hung cabinets and storage shelving (plus contents)	Any direction	0.30	0.50
5. Suspended ceiling framing systems (See Section T22-94491 (e))	Any direction	0.30^k	0.50^k
6. Suspended or surface-mounted light fixtures[l]	Any direction	1.00	1.00
7. Power-cable driven elevators or hydraulic elevators with lifts over 5 ft: a. Hoist way structural framing providing the supports for guide rail brackets	See Part 7, Title 24	See Part 7, Title 24 and Footnote[m]	See Part 7, Title 24 and Footnote[m]
b. Guide rails and guide rail brackets			
c. Car and counterweight guiding members			
d. Driving machinery, pump unit tanks operating devices and control equipment cabinets		See Part 7, Title 24	See Part 7, Title 24

supporting base or width between the exterior legs is equal to or greater than two-thirds the height.

In addition to gravity loads such storage racks or cabinets shall be designed and constructed to resist the horizontal force F_p with the base assumed to be anchored.

[i] Mobile storage racks or cabinets mounted on wheels and not restrained by fixed tracks are not subject to approval by the department. Movable storage racks or cabinets mounted on wheels or glides restrained by fixed tracks shall be designed and constructed to resist the horizontal force F_p with the base of the rack or cabinet assumed to be anchored. Provision shall be made to resist translation perpendicular to the track and overturning both perpendicular and parallel to the track.

[j] In lieu of the tabulated values, steel storage racks may be designed in accordance with UBC standard 27–11.

[k] Ceiling weight shall include all light fixtures and other equipment or components that are laterally supported by the ceiling. For purposes of determining the lateral force, a ceiling weight of not less than 4 lb/ft² shall be used.

[l] Suspension systems for light fixtures that have passed shaking table tests approved by the Office of the State Architect or that, as installed, are free to swing a minimum of 45° from the vertical in all directions without contacting obstructions shall be assumed to comply with the lateral force requirements of Section T22-94215 (m).

Unless of the cable-type, free-swinging suspension systems shall have a safety wire or cable attached to the fixture and structure at each support capable of supporting 4 times the supported load.

[m] The design of guide rail support bracket fastenings and the supporting structural framing shall be in accordance with Section 3030 (k), Part 7, Title 24, using the weight of the counterweight or maximum weight of the car plus not less than 40 percent of its rated load. The seismic forces shall be assumed to be distributed one-third to the top guiding members and two-thirds to the bottom guiding members of cars and counterweights unless other substantiating data are provided.

Retainer plates are required for both car and counterweight, designed in accordance with Section 3032 (c) (1), Part 7, Title 24. Retainer plates are required top and bottom of car and counterweight except where safety devices are provided.

The design of car and counterweight guide rails for seismic forces shall be based on the following requirements.

a. The lateral forces shall be based on horizontal accelerations of 0.3g for buildings with an $I = 1.0$ and 0.5g for buildings with an $I = 1.5$.

b. W_p shall equal the weight of the counterweight or the maximum weight of the car plus not less than 40 percent of its rated load.

c. With the car or counterweight located in their most adverse position, the deflection of the rail relative to its supports shall not exceed ½ in, and the stresses shall not exceed the limitations specified in these regulations.

Exception: Where guide rails are continuous over two or more supports and rail joints are located within 2 ft of their supporting brackets, the maximum weight of car and its rated load may be determined from Figure 3030 F1, Part 7, Title 24, using the allowable values given in the figure divided by the importance factor I of the building.

The maximum weight of counterweight for the rail size and the bracket spacing used may be determined from Figure 3030 F3A, 3030 F3B, or 3030 F3C, Part 7, Title 24, using the allowable values given in the figures divided by the importance factor I of the building.

State of California— Building Standards— Schools

DIVISION T21.   PUBLIC WORKS

CHAPTER 1.   DEPARTMENT OF GENERAL SERVICES

SUBCHAPTER 1.   OFFICE OF STATE ARCHITECT

Group 2.   Access to Public Buildings by Physically
Handicapped Persons
Group 3.   Design Criteria, Materials and Details of
Construction

Detailed Analysis

Group 2.   Access to Public Buildings by Physically
Handicapped

Article 1.   General Requirements
Section
T21-81.   Purpose
T21-82.   Authority
T21-83.   Scope
T21-84.   Exceptions
T21-85.   Compliance Procedures

Group 3.   Design Criteria, Materials and Details of
Construction

Article 1.   Administration
Section
T21-202.   Standard Reference Documents
Table T21-1-A.   List of Standard Reference Docu-
ments

Article 2.   (No Requirements)

Article 3.   General Design Requirements
Section
T21-2301.   Scope
T21-2303.   Design Methods
T21-2304.   Floor Designs
T21-2305.   Roof Design
T21-2307.   Deflection
T21-2308.   Special Design
T21-2309.   Exterior Walls
T21-2312.   Earthquake Regulations

T21-2312.   Earthquake Regulations. (a) General.
Every school building or structure and every
portion thereof including the nonstructural
components shall be designed and constructed to
resist stresses produced by lateral forces as provided
in this section. Stresses shall be calculated as the
effect of a force applied horizontally at each floor or
roof level above the base. The force shall be
assumed to come from any horizontal direction.

Structural concepts other than set forth in this
section may be approved by the office when
evidence is submitted showing that equivalent
ductility and energy absorption are provided.

When the design of a structure or parts of a structure result in unusual configuration or irregular distribution of lateral stiffness, evidence shall be presented to show that equivalent safety to that established by these regulations is provided or the office shall withhold its approval.

Where prescribed wind loads produce higher stresses, such loads shall be used in lieu of the loads resulting from earthquake forces.

Where buildings provide lateral support for walls retaining earth, and the exterior grades on opposite sides of the building differ by more than 6 ft, the load combination of the dynamic increment of active earth pressure due to earthquake acting on the higher side, as determined by a civil engineer qualified in soils engineering, plus the difference in active earth pressures shall be added to the lateral forces provided in this section. In lieu of a determination by a civil engineer qualified in soils engineering, the dynamic increment of active earth pressure for soils of low cohesion may be considered to be a uniformly distributed pressure, P_d, determined in accordance with the following formula:

$$P_d = Z(0.03\delta h)$$

where P_2 = dynamic increment of active earth pressure due to earthquake, lb/ft²

δ = soil density, lb/ft³

h = height of retained earth above bottom lateral support for wall

Z = seismic hazard zone coefficient

Table T21-23-J-PART B   Horizontal Force Factor C_p for Anchorage of Nonstructural Components*

Category	Direction of Force	Value of C_p†
1. When connected to, part of, or housed within a building:		
a. All mechanical and electrical equipment or machinery	Any direction	0.30
b. Tanks (plus contents) and support systems		
2. Emergency power supply systems	Any direction	0.75
3. a. Storage racks with the upper storage level more than 5 ft in height (plus contents)	Any direction	0.30
b. Floor-supported cabinets and bookstacks more than 5 ft in height (plus contents)		
4. Wall hung cabinets and storage shelving (plus contents)	Any direction	0.30
5. Suspended ceiling framing systems (See Section T21-4701(e))	Any direction	0.30
6. Suspended light fixtures	Any direction	1.00
7. Power-cable driven elevators or hydraulic elevators with lifts over 5 ft:		
a. Car and counterweight guides, guide rails, and supporting brackets and framing	See Part 7, Title 24	See Part 7, Title 24
b. Driving machinery, operating devices, and control equipment		

* N.B. See Section T21-2312(m) for use of C_p for design of anchorage of non-structural components in accordance with the following formula:

$$F_p = ZC_pW_p \qquad (12.9)$$

Welded, bolted, or other intermittent connections such as inserts shall not be allowed the one-third increase in allowable stress permitted in UBC Section 2303(d).

† C_p may be two-thirds of value shown for components mounted on foundations at grade or on floor slabs on earth subgrade. C_p shall be not less than the ratio of F_x/W_x for floor or roof level under consideration. Where a dynamic analysis is used in the design of the building, the forces so determined may be used in the design of the elements or components with appropriate resistance criteria.

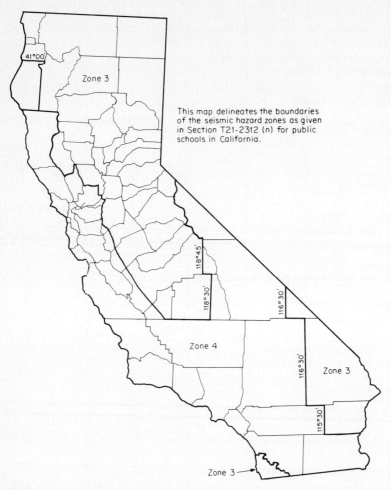

This map delineates the boundaries
of the seismic hazard zones as given
in Section T21-2312 (n) for public
schools in California.

FIG. G.1 Seismic hazard zone map for public schools in California. This map delineates the boundaries of the seismic hazard zones as given in Section T21-2312(n) for public schools in California.

SMACNA
Standard Details

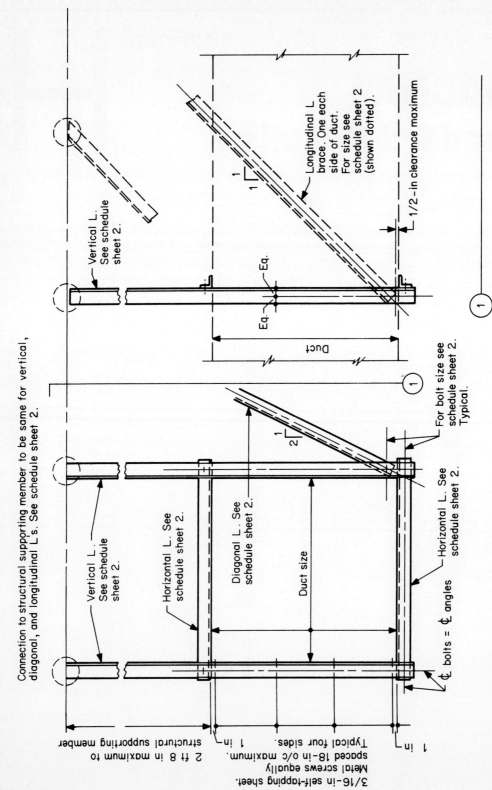

Connection to structural supporting member to be same for vertical, diagonal, and longitudinal L's. See schedule sheet 2.

Vertical L. See schedule sheet 2.

Vertical L. See schedule sheet 2.

Horizontal L. See schedule sheet 2.

Diagonal L. See schedule sheet 2.

Duct size

Horizontal L. See schedule sheet 2.

¢ bolts = ¢ angles

For bolt size see schedule sheet 2. Typical.

Vertical L. See schedule sheet 2.

Longitudinal L brace. One each side of duct. For size see schedule sheet 2 (shown dotted).

1/2 - in clearance maximum

Eq.

Eq.

Duct

①

①

①

Note!!
1. For spacing of bracing refer to typical note 3, sheet 1.
2. When a combination of ducts are used in lieu of one duct, at least two sides of each duct must be connected to vertical or horizontal L's. (Add horizontal L's if required).

2 ft 8 in maximum to structural supporting member

3/16-in self-tapping sheet. Metal screws equally spaced 18-in o/c maximum. Typical four sides.

1 in

1 in

FIG. H.1 Typical side bracing for rectangular ducts.

Connection to structural supporting member to be same for vertical, diagonal, and longitudinal L's. Use type I at 28-in dia., type II at 36-in dia. duct. See schedule sheet 9.

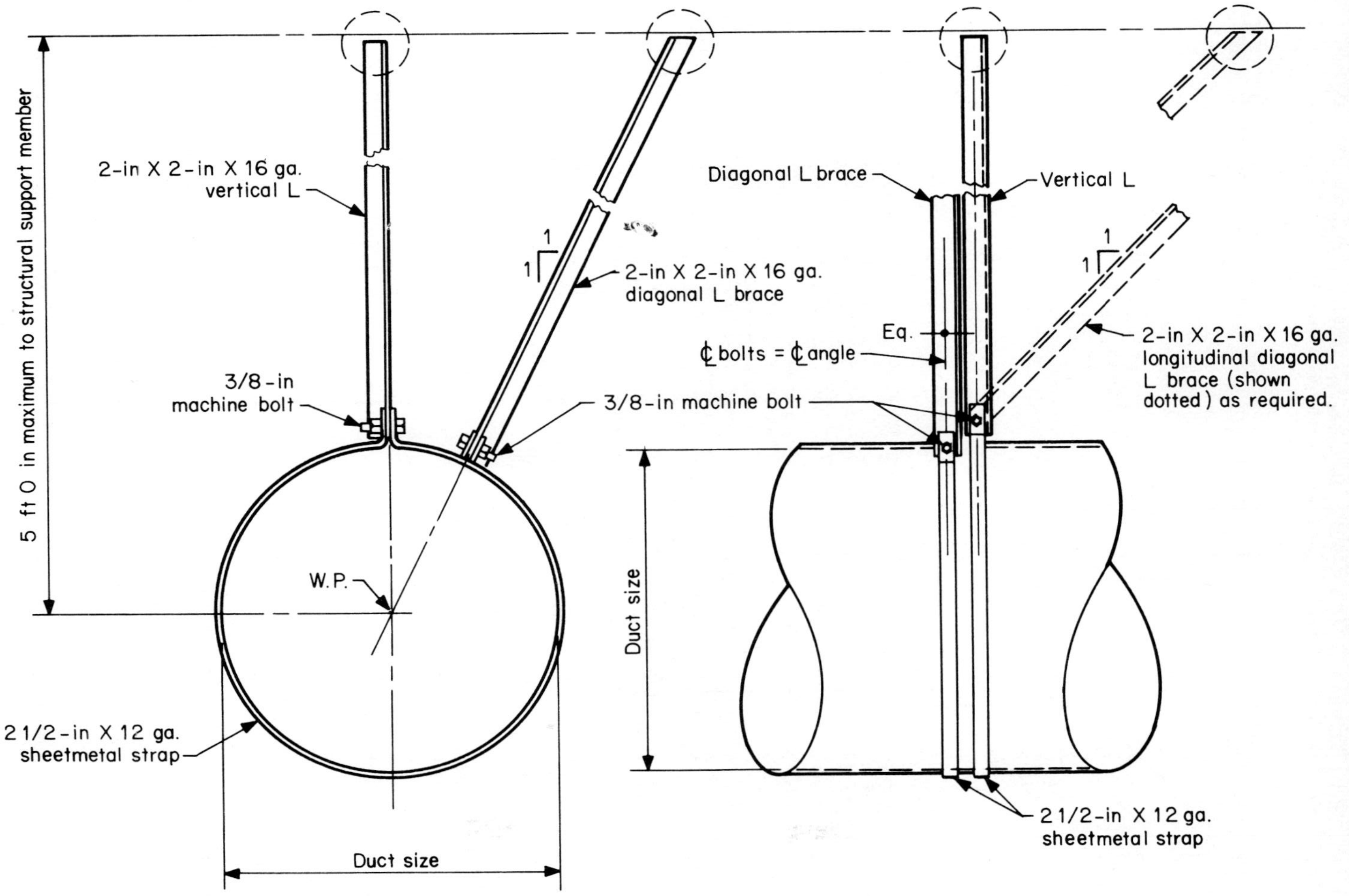

2-in X 2-in X16 ga. longitudinal diagonal L brace (shown dotted) as required.

Vertical L

Diagonal L brace

2-in X 2-in X 16 ga. diagonal L brace

Eq.

2 1/2-in X 12 ga. sheetmetal strap

$\mathcal{C}$ bolts = $\mathcal{C}$ angle

3/8-in machine bolt

Duct size

2-in X 2-in X 16 ga. vertical L

3/8-in machine bolt

W.P.

Duct size

5 ft 0 in maximum to structural support member

2 1/2-in X 12 ga. sheetmetal strap

Note II: For spacing of bracing refer to typical note 3, sheet 1.

FIG. H.2 Typical bracing for 28- and 36-in-diameter ducts.

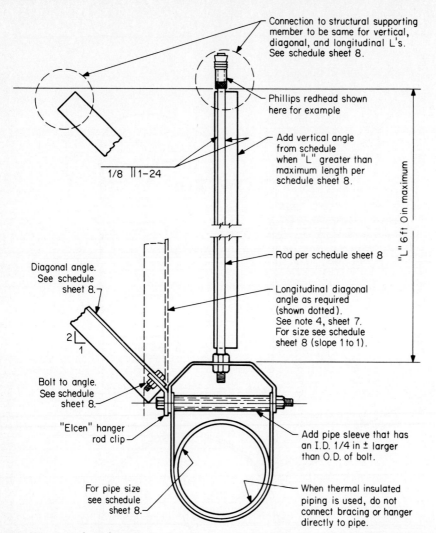

Connection to structural supporting member to be same for vertical, diagonal, and longitudinal L's. See schedule sheet 8.

Phillips redhead shown here for example

Add vertical angle from schedule when "L" greater than maximum length per schedule sheet 8.

1/8 ‖ 1-24

Rod per schedule sheet 8

Longitudinal diagonal angle as required (shown dotted). See note 4, sheet 7. For size see schedule sheet 8 (slope 1 to 1).

"L" 6 ft 0 in maximum

Diagonal angle. See schedule sheet 8.

2
1

Bolt to angle. See schedule sheet 8.

"Elcen" hanger rod clip

For pipe size see schedule sheet 8.

Add pipe sleeve that has an I.D. 1/4 in ± larger than O.D. of bolt.

When thermal insulated piping is used, do not connect bracing or hanger directly to pipe.

FIG. H.3 Typical pipe bracing.

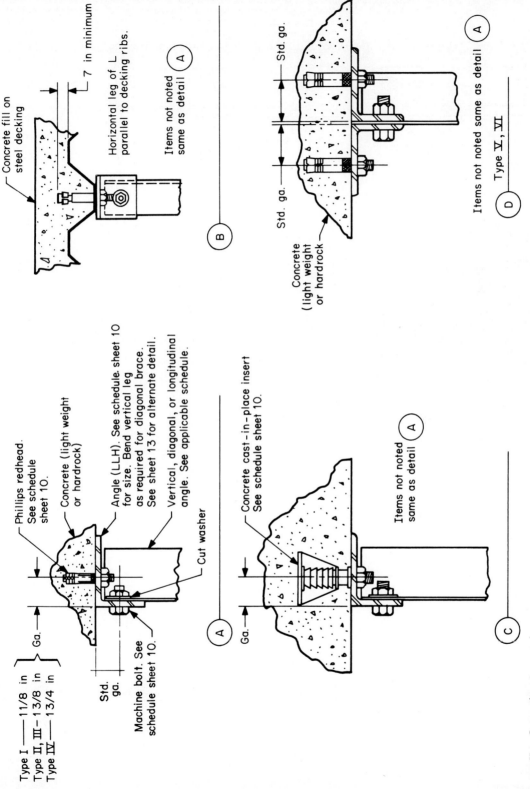

FIG. H.4 Connections to concrete.

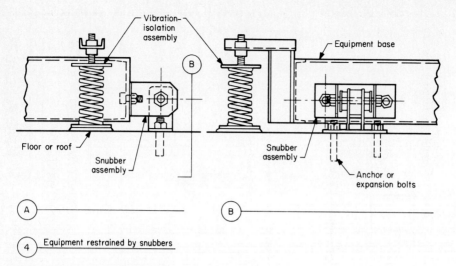

Notes:

1. Snubbers shall be designed to withstand 1.0 g (0.4 g)* lateral and vertical loads.
2. Snubbers shall be series Z-1011 as manufactured by Mason Industries, Inc., or approved equal.
3. Install four all-directional snubbers that are double-acting and located as close to the vibration isolators as possible.
4. Snubbers shall consist of interlocking steel members restrained by 3/4-in minimum removable neoprene compounded to bridge bearing specifications.
5. Snubbers shall be manufactured with an air gap between steel and neoprene of not less than 1/8 in nor more than 1/4 in.

* Same as note 3 of detail ①

FIG. H.5.   Floor- or roof-mounted equipment with vibration isolation continued.

Seismic Zone Maps*

* The maps in this appendix are courtesy of Wyle Laboratories.

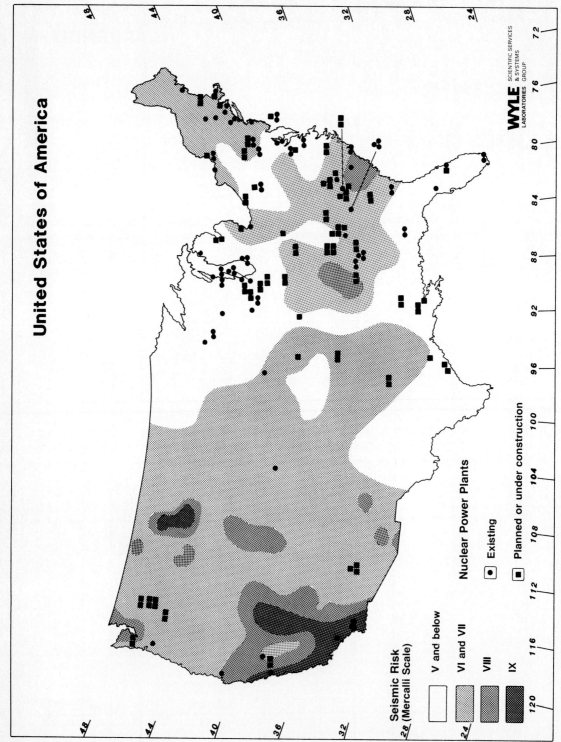

FIG. I.1   Seismic zone map of the United States. *(Wyle Laboratories.)*

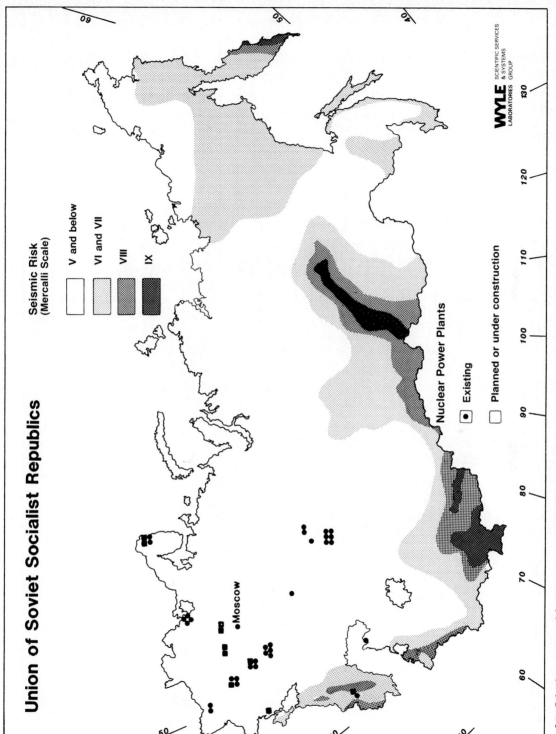

FIG. I.2   Seismic zone map of the Soviet Union. (*Wyle Laboratories.*)

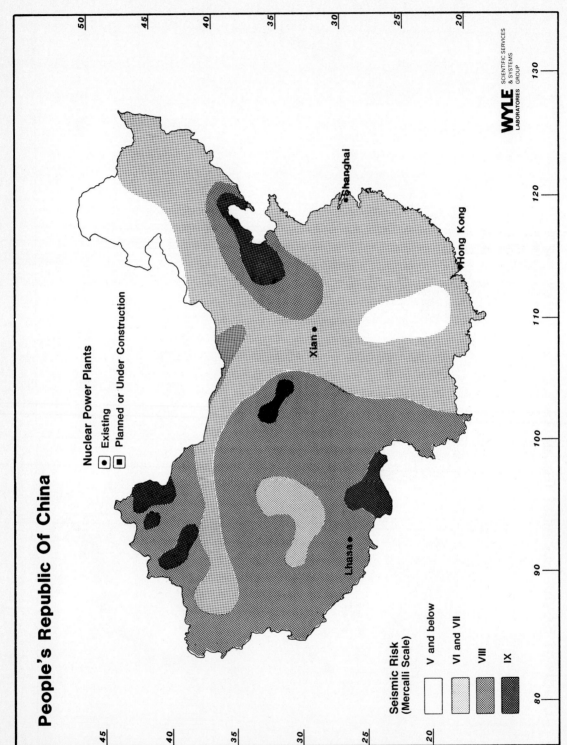

FIG. I.3   Seismic zone map of the People's Republic of China. *(Wyle Laboratories.)*

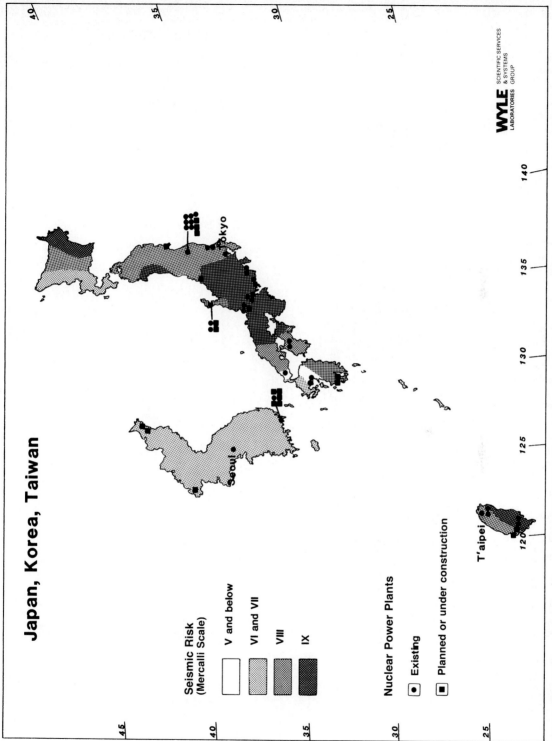

FIG. I.4 Seismic zone map of Japan, Korea, and Taiwan. (*Wyle Laboratories.*)

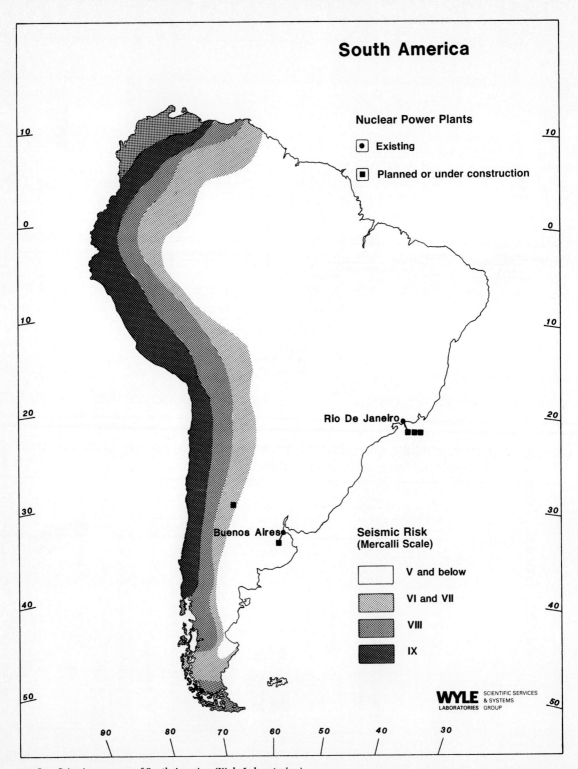

FIG. I.5 Seismic zone map of South America. *(Wyle Laboratories.)*

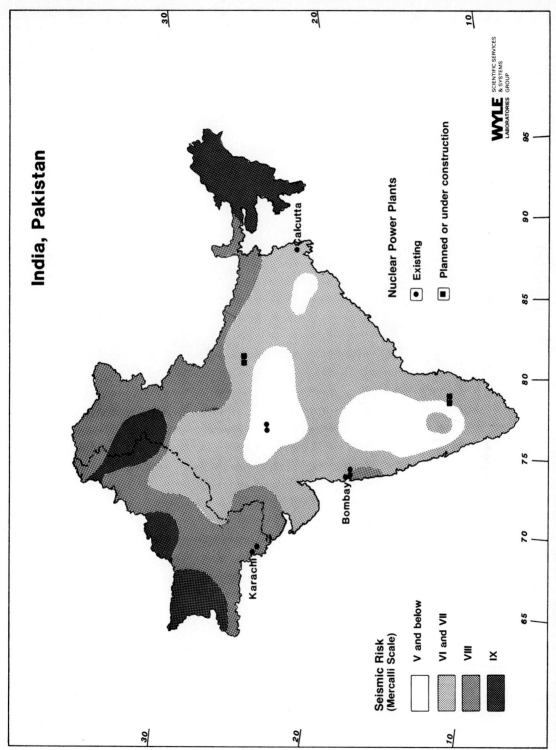

FIG. I.6   Seismic zone map of India and Pakistan. *(Wyle Laboratories.)*

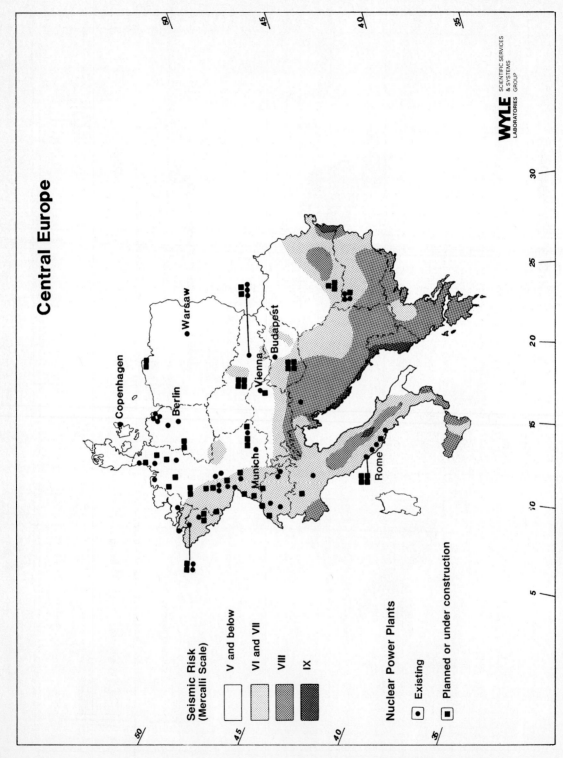

FIG. I.7   Seismic zone map of central Europe. (*Wyle Laboratories.*)

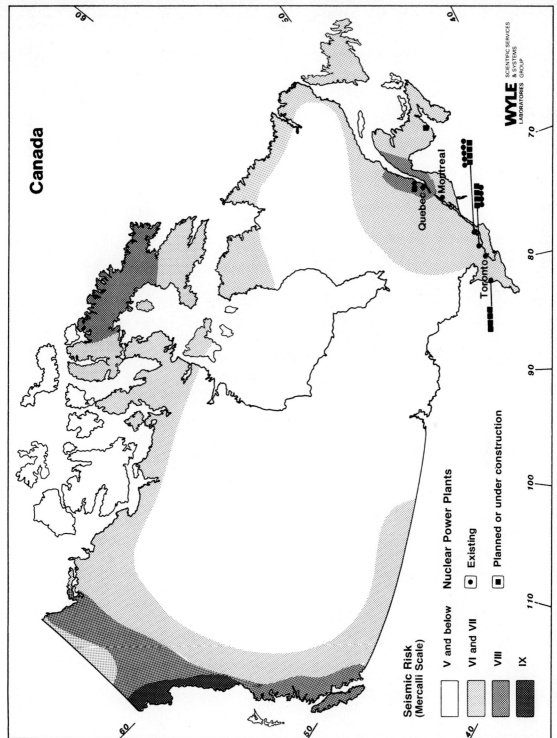

FIG. I.8   Seismic zone map of Canada. *(Wyle Laboratories.)*

Canada

Seismic Risk
(Mercalli Scale)

V and below

VI and VII

VIII

IX

Nuclear Power Plants

● Existing

■ Planned or under construction

Quebec

Montreal

Toronto

WYLE LABORATORIES
SCIENTIFIC SERVICES
& SYSTEMS
GROUP

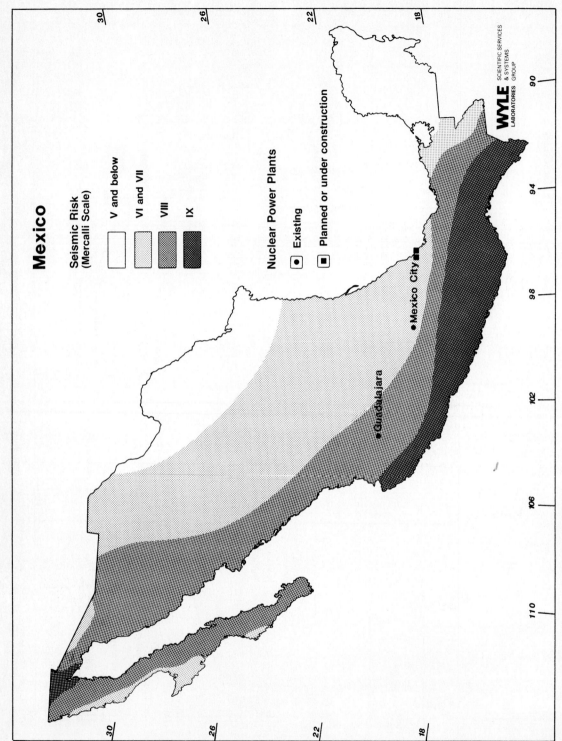

FIG. I.9   Seismic zone map of Mexico. *(Wyle Laboratories.)*

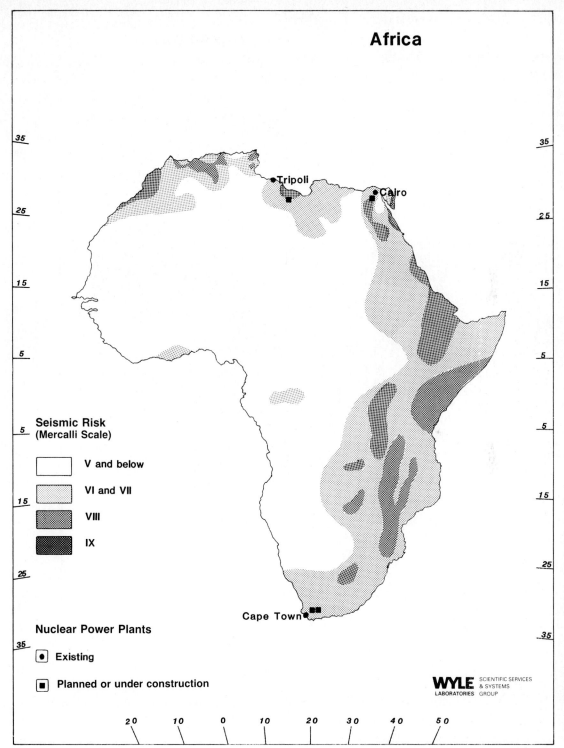

Africa

Seismic Risk
(Mercalli Scale)

V and below

VI and VII

VIII

IX

● Tripoli

● Cairo

Cape Town ●

Nuclear Power Plants

● Existing

■ Planned or under construction

WYLE SCIENTIFIC SERVICES
LABORATORIES & SYSTEMS
GROUP

FIG. I.10   Seismic zone map of Africa. *(Wyle Laboratories.)*

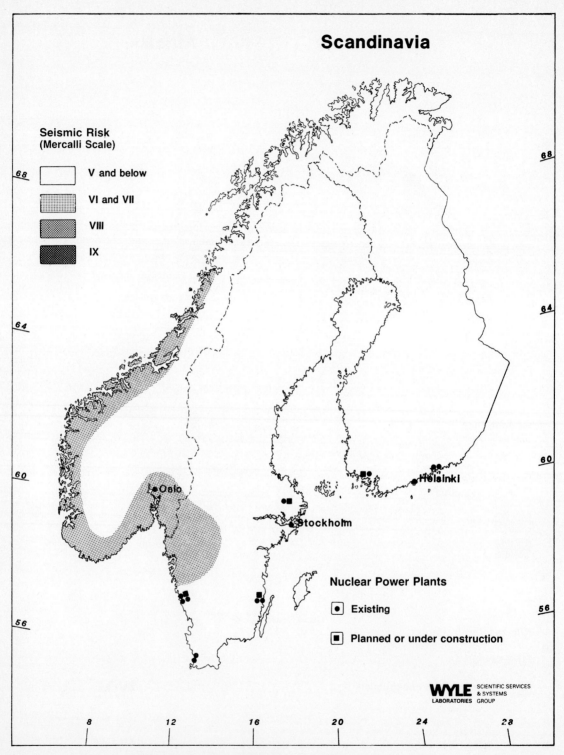

FIG. I.11   Seismic zone map of Scandinavia. *(Wyle Laboratories.)*

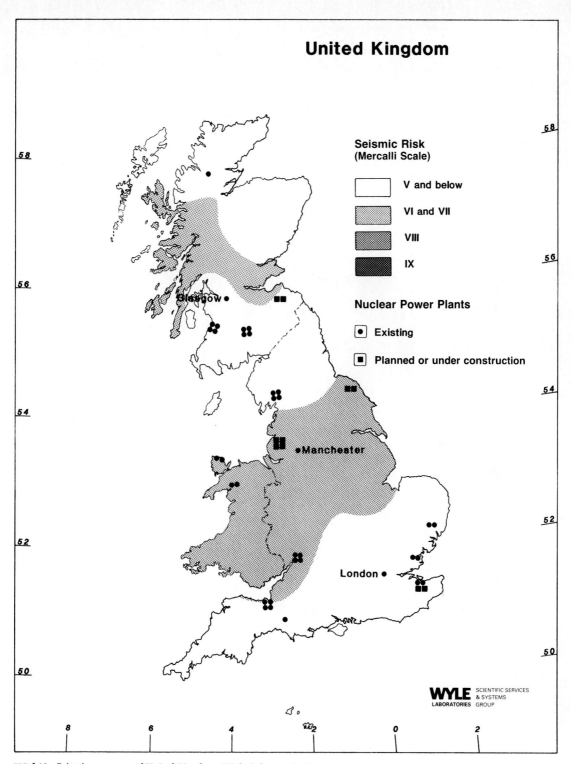

FIG. I.12   Seismic zone map of United Kingdom. *(Wyle Laboratories.)*

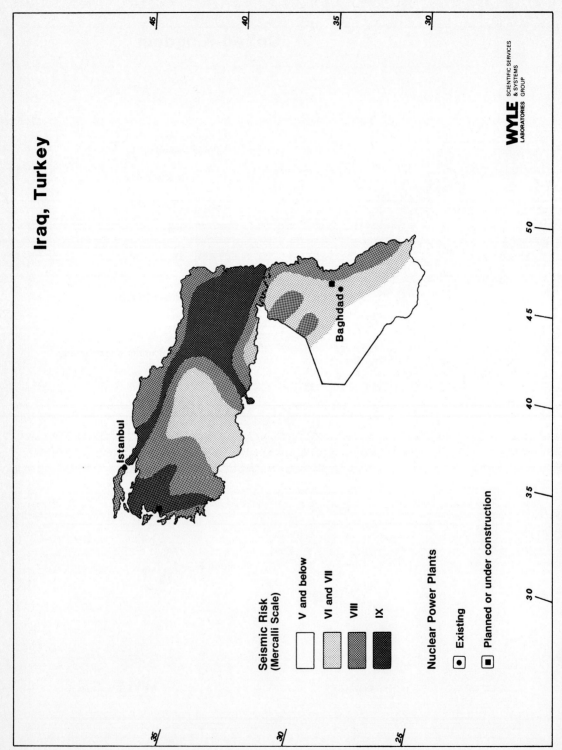

Iraq, Turkey

Istanbul

Baghdad

Seismic Risk
(Mercalli Scale)

V and below

VI and VII

VIII

IX

Nuclear Power Plants

● Existing

■ Planned or under construction

WYLE
LABORATORIES

SCIENTIFIC SERVICES
& SYSTEMS
GROUP

FIG. I.13   Seismic zone map of Iraq and Turkey. *(Wyle Laboratories.)*

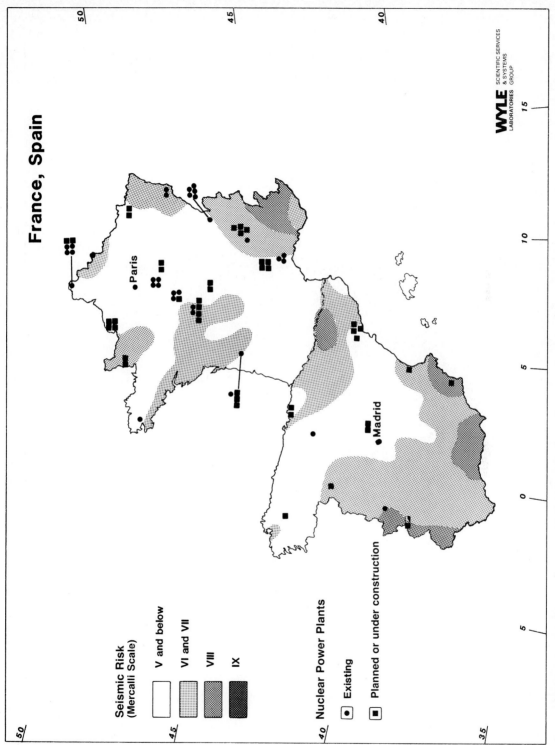

France, Spain

**Seismic Risk
(Mercalli Scale)**

V and below

VI and VII

VIII

IX

Nuclear Power Plants

● Existing

■ Planned or under construction

Paris

Madrid

WYLE
LABORATORIES

SCIENTIFIC SERVICES
& SYSTEMS
GROUP

FIG. I.14   Seismic zone map of France and Spain. (*Wyle Laboratories.*)

Bibliography

American Institute of Steel Construction Inc.: New York.

American Plywood Association: *Plywood Construction for Noise Control*, Bulletin W460 277, Tacoma, Wash., 1960.

American Society of Heating, Refrigerating and Air-Conditioning Engineers Inc.: *Fundamentals*, 1977; *Systems*, 1980, New York.

Bolt, Bruce A.: *Earthquakes—A Primer*, W. H. Freeman, San Francisco, 1978.

Concrete Reinforcing Steel Institute: Chicago, Ill.

Gypsum Association: *Fire Resistance Design Manual*, Bulletin GA-600-78 or latest edition, Evanston, Ill.

International Conference of Building Officials: *Uniform Building Code*, Whittier, Calif.

Iwan, W. D. (ed.): *Strong-Motion Earthquake Instrument Arrays*, California Institute of Technology, Pasadena.

Lead Industries Association: "Improved Sound Barriers Employing Lead," A.I.A. no. 39, *Practical Application of Sheet Lead for Noise Barriers*, Bulletin 15-M-6-73, New York.

Machinery's Handbook, 20th ed., Industrial Press Incorp., New York, 1978.

Mason Industries Inc.: *Seismic Control Specifications for Floor Mounted Equipment*, SCS-100 Bulletin, Hollis, N.Y., June 1975.

Office of Architecture and Construction: Register 71, no. 52, 12-25-71.

Research Staff Office of Construction: *Study to Establish Seismic Protection Provisions for Furniture, Equipment and Supplies for VA Hospitals*, Veterans Administration, Washington, January 1976.

Sheet Metal and Air-Conditioning Contractors National Assoc. Inc.: *Guidelines for Seismic Restraints of Mechanical Systems*, Metal Industry Fund of Los Angeles, Los Angeles, March 1976.

Sheet Metal and Air-Conditioning Contractors National Assoc. Inc.: *Duct Liner Applications Standard*, Tysons Corner, Vienna, Va., 1975.

United McGill Corporation: Bulletin no. 15-1-776, 1971, Bulletin no. 30-1-276, 1975, Bulletin no. F.O. 40-1-472, 1969, Westerville, Ohio.

Wire Rope Technical Board: Bethesda, Md.

Index

This index consists of two parts: a subject index and a geographical index.

Subject Index

Absorption, sound, 168
Absorption units, steam, 137–138
Access doors, 78–79
Acoustical enclosures, 14, 58, 78, 130–133,
    149–160, 164–170
Air bag, 52
Air-handling units, 47, 48, 144–159, 294
Air springs (mounts), 52–56, 286–289, 293
Alignment, 24, 33, 36, 92, 96, 221
All-directional snubbers, 311–313
Anchors, resilient pipe, 301
Anticavitation devices, 110
Attenuation path, noise, 134
Axial fans, 49, 143, 305

Barriers (see Acoustical enclosures)
Base frames, 171–173
Bellows, metal corrugated, 124, 190
Beverage coolers, 336–340
Blowers, 261, 263, 265, 267
Boilers, 119–121, 260

Bolts:
    pull-out data for, 314
    sizing, 281, 285
Brackets, cantilevered, 44, 46, 53, 185, 308
Breeching, 119–121
Building standards, California, 392–398
Bus bars, 295

Cable trays or racks, 210
Cables:
    electrical, 202, 203, 205
    steel, 269, 295–305
Cantilevered brackets (see Brackets)
Captive isolators, 291
Casing radiation, 64, 149
Cavitation, 107–110
Ceilings, 227, 242–248
Check valves, silent, 107, 109
Chillers, gas-engine-driven, 284, 314, 315
    (See also Cooling towers)
City water connections, 119

Collapse, spring, 25, 39, 91
Commercial buildings, seismic snubber analysis for, 262–267
Compressors:
　air, 122–124, 261, 263, 265, 267, 292
　reciprocating, 122
　refrigeration, 336
Concrete inertia bases, 97, 139, 171–175, 284
Condensate pumps, 137
Condensing units, 164–170
Conductors, electrical, 202–205, 209, 216, 217, 219
Conduits, electrical, 136, 201, 205-209
　multiple, 204, 211–212
Control rods and cables, 98
Conversion, SPL to dB(A), 347
Conveyors, vibrating, 319–320
Cooling towers, 36, 40, 41, 81, 116, 125–136, 302
　(See also Chillers)
Cork pads, 148
Corrugated deck, sealing, 230
Curbs, vibration-isolation, 12, 166, 167
Cushion clamps, 189

Debris, removal of, 138
Decibel, 3, 347
Decibel addition, 347–348
Deflection, 8, 10–11, 15
Diesel generators, 63
Dimmers, 222–223
Door gaskets, 249–251
Doors, 78–79, 249–251
Drain piping, 138
Duct fittings, 66
Duct isolation, 66, 75, 242
Duct lining, 14
Ducts, 16
　penetrations and connections, 70–79
　silencers for 57–69
　supports for, 76–77
Dunnage frames, 127, 135
Dynamic analysis, 259, 268, 269
Dynamic insertion loss (DIL), 3, 14, 67, 69

Earthquakes, 255
　zone maps, 256, 282, 405–419
　(See also Seismic isolation and protection)

Ejector pumps, 196
Elbows:
　duct, 64
　pipe support, 97
Electrical connections, 119, 124, 136
　distribution panels, 206
Elevator machinery, 322–328
Emergency electric generators, 220–221, 257
Enclosures:
　acoustical (see Acoustical enclosures)
　gypsum board, 60, 157, 160, 162
　lead sheet, 154, 155
　leaded vinyl, 105, 106, 154
　loaded vinyl, 59
　plywood, 150–152, 160
　sheet metal, 153
Equipment bases, 11–12
Escalators, 341–347

Fan coil units, 142
Fans, 49, 143, 161–162, 305
Fasteners, threaded, rivets and, 295, 365–374
Felt padding, 184
Fiberglass, 132, 133, 155–157, 231–233, 236–238, 242–246
Fire dampers, 61, 62
Flanking, 228
Flexible connections, 116–117, 142, 143, 269, 284
　boiler breeching, 120–121
　conduits, 207–209, 214–215, 219, 223
　ducts, 74, 143
　piping, 98, 190–192
Floating floors, 227, 236–241
Floor drains, resilient, 240–241
Floor-mounted equipment, 9, 49, 171–175
Floor-mounted piping, 15, 51, 80, 84, 85, 88–89
Foam rubber, 32, 73, 211–212, 219
Frequency, 3

Gas connections, 119–120
Gaskets, 187, 249–251
Generators, electric, 220–221, 257
Grommets, resilient, 37, 42, 193, 206, 310
Guidelines, 7–17
Gypsum board, 59, 230, 244, 245
　for enclosures, 60, 157, 160, 162

Hangers, 10, 21–27, 90–96, 111–115, 310
    alignment of, 21, 25, 26
    box rotation, 21–23, 141
    brackets for, 114–115
    inertia, 247–248
    overload, 23, 25
    placement of, 21–22, 27, 95
    precompressed, 111
    rods for, 22, 26, 91, 92
Health facilities, California building standards for, 392–398
    (See also Hospitals)
Heaters, unit, 163
Heating and ventilating (H&V) units, 260, 262, 264, 266
Hertz, 3
Hold-down bolts, 34, 36–37, 40, 42, 145, 147, 293, 420
Hospitals:
    safety of construction of, 388–391
    seismic snubber analysis, for, 260–261, 264–267
    (See also Health facilities)
Housed springs, 28–35, 221
Housekeeping pads, 54, 139
Hubless fittings, 187–188
Hydraulic piping, 184, 185, 193, 322–325

Inertia bases, concrete, 97, 139, 171–175, 284
Inertia hangers, 247–248
Insertion loss:
    dynamic, 3, 14, 67, 69
    static, 67
Inspection, 26, 34
Installation, 15, 141
Internal duct lining, 14
Isolators (see specific isolators)

Jack-up isolators, 238

Laboratory equipment, 76
Lead sheet, 154, 155
Leaded vinyl, 59, 105, 106, 154
Lining, acoustical duct, 14
Loaded vinyl, 59
Lockout system for seismic control, 276–280

Metal sandwich panels, 131
Misaligned isolators, 32, 33, 43
Misalignment, 25, 26, 93
Motor enclosures, 153–155, 158, 159
Motor generator sets for elevator, 326
Muffling orifices, 102, 168
Mylar, 132

Neoprene elements, 22, 55, 117, 141, 187, 193, 206
Neoprene pad isolators, 9, 129, 148, 180, 195, 197, 214–215, 217, 222–223, 307, 326, 327
Noise criteria (NC) levels, 3, 4, 125, 150–152, 329, 336–337, 345
Noise sources and paths, 2–3, 118, 134, 164, 227
Nuclear plants:
    seismic restraints in, 294
    vibration isolation in, 294

Open spring isolators, 35–45, 220, 308
Outdoor installation of metal parts, 146
Overextension, spring, 35
Overloading:
    neoprene elements, 25
    neoprene pads, 218
    springs, 25, 30, 39, 40, 94, 96

Packing, 82–83, 97
Pads (see Neoprene pad isolators)
Panels, acoustical, 131, 164–165, 168–170
Paths (see Noise sources and paths)
Penetrations:
    duct and pipe, 59, 70–71, 97–98, 118
    of structure, 16–17, 82–83, 182
Perforated metal panels, 132
Pipes, 179–189
    multiple, 111–113, 304
    single, 80
Piping, 322–325
    bases for, 97
    cable restraints for, 295–304
    cast-iron, 183
    chilled water, 186
    condenser water, 80, 81, 84, 85, 116
    drain, 126, 138, 173
    pulsation, 107–110

Piping (*Cont.*):
  rigid support of, 15
  sanitary, 97, 98, 182, 185
  steam, 99, 100–106
Platform (position) hangers, 310
Plumbing pipes, 179–189
Plumbing pumps, 190–197
Plywood enclosures, 150–152, 160
Pneumatic mounts, 53–54
Polyethylene, 132
Polyurethane, 346
Precompressed hangers, 111
Pressure-reducing stations, valve and pipe
    sizes for, 99–105
Product data, 9
Pulsation dampers, 107–110
Pulsatrol, 107, 109
Pumps, 190–197, 260, 262, 264, 266

Quality assurance, 8

Radiation, casing, 64, 149
Rain hood, 72–73
Receiver, sound, 2, 164
Refrigeration machines, 260, 262, 264,
    266
  steam absorption, 137–138
Resilient fittings, 197, 211, 221, 228, 240–
    243, 344
  (*See also* Snubbers, resilient)
Restraints, seismic, 257, 271–280
Richter scale, 256, 268, 280
Rivets and threaded fasteners, 295, 365–
    374
Rods, hanger, 22, 26, 91, 92
Roof drains, resilient fittings for, 240–
    241
Roof leaders, 183
Rooftop equipment, 49, 145–147, 161,
    164–170, 260, 262, 264, 266
Rotation, horizontal axis, 293
Rubber, 55, 211

Sanitary piping, 97, 98, 182, 185
Schools, California building standards for,
    396–398
Sealing, acoustical, 15, 201

Seismic isolation and protection, 243,
    254–315
  excerpts from Uniform Building Code,
    379–387
  restraining devices, 257, 271–280
  selection tables, 260–267
  snubbers (*see* Snubbers, seismic)
  zone maps, 256, 282, 405–419
Self-noise (*see* Dynamic insertion loss)
Sewage pumps, 193
Shock absorption, 107
Short circuits, vibration-isolation, 86–87
Silencers, 57–69
  elbow, 61
  pipe, 102–105
  rectangular, 57, 59, 63–65, 134
  round or conical, 64, 65, 153, 155
  specifications for, 13–14
  square, 58
Sizing (*see* Seismic isolation and protec-
    tion)
Sleeves:
  conduit, 211
  pipe, 11, 82–83, 90
  wall, 11, 90
SMACNA (Sheet Metal and Air-Condition-
    ing Contractors National Assoc. Inc.)
    standard details, 399–404
Smith-Emery test, 375–378
Snubbers:
  resilient, 45–46, 140, 144, 221, 309, 344
  seismic, 54, 257–259, 269, 271–280, 285,
    306, 311–313
  load test on, 375–378
Soil pipe, 183–184, 186–189
Sound:
  absorption of, 168
  loudness, pitch, and noise, 2, 4, 5
  (*See also* Noise sources and paths)
Sound power level (PWL), 4
Sound pressure level (SPL), 5, 347
Special equipment, 317–347
Specifications, 8–17
Spectrum, sound, 5
Springs:
  air, 52–56, 286–289
  housed, 28–35, 221
  overloading (*see* Overloading, springs)
  steel, 308
  unhoused, 35–41, 44–46

Stage house roof, ducts on, 161–162
Stage lift machinery, 321
Stanchions, 46, 51, 85, 319, 320
Static analysis, 259, 270
STC (sound transmission class), 228, 236
Steam, 99, 100–106, 137–138
Stic-klips, 156
Stiffness ranges for springs, 9
Structural separation, 234
Structural supports, 49
Styrofoam, 234
Submersible pumps, 190–191
Submittals, equipment and materials, 8–9
Sump pumps, 193–194, 196, 197
Supervision, 34, 45, 47, 85, 95, 149, 154
Suppressors, pipe noise, 102–105

Thimbles, 297–299, 305
Thrust restraints, 74
Toilet fixtures, isolation for, 181
Torque, starting, 46
Transformers, 213–219, 288–290, 310
  floor-mounted, 214–215
  hung, 213
  portable, 218
Transmission loss (TL), 3, 150, 335
Trapeze hangers, 111–115
Travel limit stops, 10, 36, 40–42, 126–128,
  146, 290, 307
Tripod isolators, 307
Turnbuckles, 295, 304

U-bolts, 180, 324–325
Uniform Building Code, 379–387
  tables, 385–387, 393
Unit heaters, 163
Utilities in wall, 235

Velcro, 154
Vertical pipe supports, 83
Vibrating conveyors, 319–320
Vibration acceleration levels, 346, 348
Vibration displacement, 16
Vibration-isolation unit:
  adjustment of, 30, 35, 49
  bolting: resilient, 36–37, 42
    rigid, 34, 40
Vinyl:
  leaded, 59, 105, 106
  loaded, 59

Walls, 227–235
Washers, resilient, 37, 42, 193
Waste, 187
Water hammer, 107
Waterproofing, 49
Weathercap, 72–73
Weatherproofing, 49
Weight distribution, 120, 128
Weld nail pins, 154–156
Winch blocks, head and loft, 334–335
Winch control console, 331
Winches, 329–335
  banner, 331–333
  batten, 329–331
  spot-line, 332–334
Wire braid, 124
Wire rope, 295–305, 349–364
Wire rope assemblies, 363–364
Wire rope clips, 298, 355–357
Wrapping, pipeline, 105–106

X-Y-Z axis, 299, 305

Zone maps, seismic, 256, 282, 405–419

Geographical Index

Australia, 47, 48, 210, 218
   Melbourne, 186
   Sydney, 139

Bikini, 256
Boston, Massachusetts, 135

California, 256, 270
   Calaxio, 268
   Los Angeles, 256
   San Andreas, 256
   Santa Barbara, 268
Canada, 66, 221
   Halifax, Nova Scotia, 26
   Montreal, 144
   Quebec, 31
   Toronto, 49, 61, 118

Detroit, Michigan, 44

Georgia, 38

Hong Kong, 47

Kansas City, Kansas, 131

Louisiana, 28
   New Orleans, 230

Maine, 164
Manila, Philippines, 46
Maryland, 49
   Baltimore, 210

New Hampshire, 141

Puerto Rico, 49, 161

Singapore, 125, 129

Venezuela, 185
   Caracas, 228

About the Author

Robert S. Jones is a draftsman, designer, engineer, and acoustical consultant. With on-site experience at building projects around the world, he is the author of more than 20 articles and lectures on noise and vibration control.